The Basics of Chemistry

The Basics of Chemistry

Ashvani Kaushik

ANMOL PUBLICATIONS PVT. LTD.

NEW DELHI - 110 002 (INDIA)

ANMOL PUBLICATIONS PVT. LTD.
H.O.: 4374/4B, Ansari Road, Darya Ganj,
New Delhi-110 002 (India)
Ph.: 23278000, 23261597
B.O.: No. 1015, Ist Main Road, BSK IIIrd Stage
IIIrd Phase, IIIrd Block,
Bangalore - 560 085 (India)
Visit us at: www.anmolpublications.com

The Basics of Chemistry

First Published, 2007
ISBN 978-81-261-3410-6

PRINTED IN INDIA

Printed at Mehra Offset Press, Delhi.

Contents

Preface

The body of knowledge called Science consists primarily of models and concepts, based on observations and deduced from careful reasoning. Viewed in this way, Science is a creative human endeavor. The models, concepts, and theories we use to describe nature are accomplishments equal in creativity to any artistic, musical, or literary work.

Unfortunately, textbooks in chemistry traditionally present these models and concepts essentially as established facts, stripped of the clever experiments and logical analyses which give them their human essence.

As a consequence, students are typically trained to memorize and apply these models, rather than to analyse and understand them. As a result, creative, analytical students are inclined to feel that they cannot "do" chemistry, that they cannot understand the concepts, or that chemistry is dull and uninteresting. This book in Chemistry is presented to redirect the focus of learning.

We have tried to cover the basic areas of chemistry and also tried in simple language to present these topics. This book is an attempt to have a look on the basics of chemistry. We hope that this book will be very useful for the students.

Author

Chapter 1

Introduction

Matter and energy constitute the universe. Matter is said to be anything which occupies *space* and has *weight,* while energy is probably best defined by saying that it is the *ability to do work.* Just what energy really is seems at present to be beyond even our loftiest of minds. However, all of us are familiar with its common forms and manifestations. The six common forms of energy are: *heat, mechanical, electrical, chemical, light and sound.* From a practical point of view, then, anything which gives us power, and is not capable of occupying space or having weight, is energy all else is matter.

LAW OF CONSERVATION OF MATTER

Experimentally established by the great French scientist and father of chemistry, Lavoisier, the Law of Conservation of Matter states that matter is neither created nor destroyed, but may be changed from one form into another. Before Lavoisier s time this was not appreciated because scientists were misled by the phenomenon of burning. They, reasoned that the substance which burns is changed into a weightless state which escapes into the air, never to be returned.

By an ingenious experiment Lavoisier removed this fallacy, and made way for future scientists to develop the beautiful science of chemistry. He proved that when a substance burns in air it combines with oxygen and actually becomes heavier, because the substance now has the additional weight of the oxygen. This principle can be readily demonstrated by burning a piece of magnesium ribbon and then weighing the white ash which remains the ash weighs more than the original ribbon!

LAW OF CONSERVATION OF ENERGY

The Law of Conservation of Energy is identical to the above. It is common experience that energy may change from one form into another form. That the form disappears is true, but the energy which it possessed is of the same magnitude in the new form. When heat is applied to water, and the steam is made to turn a wheel, and the wheel in turn made to run a generator to produce electricity, we are passing through three forms of energy:-heat, mechanical and electrical. Theoretically, the quantity of electrical energy is equal to the quantity of heat energy.

A REVISED CONCEPT

Our atomic age has suggested some drastic changes in fundamental thinking relative to matter and energy. A concert of genius, effort and billion-volt machines has enabled man to look closer at the true "substance" of the earth and the universe. It is now a known fact that matter can be transformed into energy. This was predicted by Einstein in his now famous equation, $E = mc^2$; i.e., energy equals mass (matter) times the speed of light squared; this fact was established by the first atomic blast.

Although the Law of Conservation of Matter seems to be somewhat voided by this fact, let us not lose sight of its tremendous value to "everyday chemistry," which is what concerns us here. For all practical purposes, the law stands. It is interesting to note, however, that some scientists believe that energy and matter are one and the same. They look upon matter as being concentrated energy. Although it is not possible to conceive of this in conventional terms, modern mathematics and atomic research allow considerable support. At any rate, this concept gives us a better understanding of what is implied by those scientists who make mention of a "unified theory."

CHANGES IN MATTER

Matter may undergo two kinds of changes: *physical* and *chemical.* Any change which *does not alter* the composition of a substance is called a physical change. The breaking of glass, the melting of ice, and the boiling of water are physical changes because their internal constitution remains the same.

Water is water whether it is ice or steam; and glass is glass whether it is in the window or smashed into pieces by a student who should have been studying chemistry. On the other hand, any change which *alters* the composition of a substance is a chemical change. In this type of change one or more new substances are formed from the original substance. The souring of milk, the rusting of iron, the burning of paper, and the explosion of gunpowder are examples of chemical changes. In each instance the internal composition has been altered in such a way that the original substance no longer exists.

CHEMICAL AND PHYSICAL PROPERTIES

The chemist is exact and precise in his description of matter. Once a new substance is discovered he makes a careful study of all its properties. The properties which are ascribed to pure chemical substances are of two kinds: *chemical* and *physical* A physical property is a characteristic of a substance which may be ascertained or observed without chemical action; a chemical property is a characteristic which is determined by observing its action, or lack of action, with other substances.

To illustrate, let us consider the well-known gas oxygen. Oxygen is colourless, heavier than air, and insoluble in water; these are its physical properties. The fact that it reacts with other substances to support combustion is a chemical property. Since it does burn itself, this would be a chemical property of a negative nature. Any large work on chemistry always contains a detailed account of the physical and chemical properties of important compounds.

STATES OF MATTER

Matter may exist in three states, or forms: *solid*, *liquid* and *gaseous*. A solid is characterized by its definite shape and volume; a liquid by its definite volume and readiness to assume the shape of the receptacle into which it is placed; and a gas by its indefinite shape and volume. Many substances may exist in two or all three states. Such a substance, however, retains the same composition regardless of its state. As pointed out above, water is water whether it is liquid, steam or ice. The change of state is a physical act. It is now known that the three states of matter are merely examples of an energy differential. In the solid condition the molecules have little energy and, therefore, possess little motion.

This allows for more compactness and accounts for the density of the solid state. In the gaseous state the molecules of a substance have the greatest amount of energy, and are in rapid motion. This accounts for the swift movement and diffusion of the gaseous state. The liquid state is merely an intermediate energy level. The energy differential of the three states is readily appreciated when one compares the amount of energy needed to convert ice and water to steam. Since ice, the solid state, represents a lower energy level than liquid water, it follows that more heat energy must be added to the former than the latter.

KINDS OF MATTER

First of all let us attempt to explain what we mean by the rather awkward term, substance. As may be inferred by those who have read this far, the word has been employed already many times for the want of a better noun. A glance at the dictionary will convince one of its fluid character. Perhaps it is best to say that it is synonymous with the word matter, and used frequently in this text to avoid monotony. To be sure, there are three kinds of matter (or substances): *elements*, *compounds* and *mixtures*.

ELEMENTS

An element is a substance which cannot be decomposed into simpler material by any ordinary chemical means. (Years ago an element would have been defined simply as a substance which cannot be decomposed into simpler material, with the phrase "by ordinary chemical means" omitted. Subsequently we shall see that this change was caused by our present concept of atomic structure.) The earth and its flora and fauna are constituted by 92 of these elemental substances in an astronomical number of combinations. All things which compose our material world are either elemental in nature or, as is usually the case, composed of a combination of elements. In order that the chemist may conveniently describe the constitution of matter he employs a system of chemical shorthand whereby each element is assigned a *symbol.* The symbol consists of one or two letters of the English or Latin name of the element. The following elements are important to us because they are the principal constituents of our biologic world:

Element	Symbol
Oxygen	O
Carbon	C
Hydrogen	H
Nitrogen	N
Calcium	Ca
Phosphorus	P
Potassium	K
Sulfur	S
Sodium	Na
Chlorine	Cl
Magnesium	Mg
Iron	Fe
Manganese	Mn
Copper	Cu
Iodine	I

CLASSIFICATION OF THE ELEMENTS

Elements are classified into two general classes: *metals* and *nonmetals*. Chemically, some elements exhibit both metallic and nonmetallic properties; such elements are called *amphoteric* elements. The table below presents the more characteristic properties of metals and nonmetals.

Metals	*Nonmetals*
Physical Properties Lustrous	Nonlustrous
Generally high density	Low density
Good conductors of electricity	Poor conductors
Chemical Properties	
Have an affinity for nonmetals; do not chemically combine with each other	Have an affinity for metals; do not combine with each other

COMPOUNDS

A compound is a substance consisting of two or more elements chemically united; that is, combined in such a manner that each constituent element has lost its individual characteristics. Each compound has its own static personality by virtue of another very basic law of chemistry: the *Law of Definite Proportions*. The Law of Definite Proportions states that when two or more elements chemically combine, they always do so in a *definite proportion* by weight.

If this were not true, and elements were as changeable as the weather in the proportion in which they combined with one another, the resulting compound would have variable properties. That this is not the case is most fortunate, for it would mean that when hydrogen and oxygen combine to form water in Moscow it would not necessarily be the same as water formed from a combination of hydrogen and oxygen in the city of New York. The Law of Conservation of Matter and the Law of Definite Proportions, then, render validity to the whole science of chemistry, for without matter's unyielding devotion to these two principles man would be unable to have confidence in his material world. Cane sugar, table salt and water are three common examples of true compounds.

MIXTURES

A mixture is a substance consisting of two or more elements or compounds not chemically combined and in any proportion. Anything that is not a pure compound or a pure element is a mixture by virtue of deductive reasoning. Air, soil, petroleum, candy and blood are representative examples of mixtures. A mixture may be conveniently differentiated from a compound by reference to the following table.

Mixture	Compound
1. Constituents retain individual propertiesties	1. Do not retain propertiesties.
2. No definite composition by weight.	2. Definite composition.
3. Can be separated into its constituents by mechanical means.	3. Can be separated only by chemical means.

Chapter 2

Structure and Behaviour of Matter

What would the "end product" look like if one were to take a piece of matter, say a stone, and by some means break it up into an infinite number of pieces? More specifically, is there a fundamental unit of matter? If so, what is its structure and behaviour? Man has pondered such questions for 2000 years. This chapter is concerned with the product of his efforts to supply the answers.

ATOMIC THEORY

In the year 1803, the brilliant deductions of John Dalton presented to man the first unified idea concerning the structure and behaviour of matter. He proposed his deductions after deep consideration and appreciation of the two most fundamental laws of chemistry, the laws with which we arc already familiar, the Law of Conservation of Matter and the Law of Definite Proportions.

Dalton's deductions, classically referred to as the *Atomic Theory,* make the following assumptions:

1. All matter is composed of very small particles, called atoms.
2. All the atoms of the same element are exactly alike in size, shape and weight, but differ in all respects from the atoms of every other element.
3. Atoms of most elements are able to unite with the atoms of other elements.

4. In all chemical changes atoms do not break up, but act as individual units.

As suggested in this theory, let us adopt the atom as our unit and see if we can refine the thoughts which have already been expressed. A slight glance at the second and third statements will crystallize the rather nebulous definitions which were offered for the element and compound. We said that an element was the simplest type of matter because it could not be broken down into a more simple substance by ordinary means.

Although this is a valid definition, we are now able to offer an alternate and more precise description: *An element is a substance made up of like atoms.* This explains in a very fundamental way why an element is the most elemental substance. A compound, then, may be considered in a similar way: It is a substance composed of two or more kinds of atoms in a definite proportion by weight.

The Atomic Theory has *essentially* withstood the test of time. Its most important contribution, however, is the fact that it has served as a stimulus to the fertile minds of the great atomic scientists who followed Dalton. Today we know that although the atom is a practical unit of the universe, there are still smaller particles which constitute the atom. The size of these particles is beyond the grasp of the human mind; that they exist is supported by the labours of our greatest minds in the fields of chemistry, physics and mathematics. The twentieth century concept of the atom is embodied in the *Electron Theory* of matter.

ELECTRON THEORY

This theory holds that the atom is an arrangement of particles not unlike the solar system. It explains and clarifies in an amazing manner what matter really is, and how and why it behaves as it does. The following are the cardinal concepts of this beautiful piece of human reasoning:

1. The atoms of all elements are made up of a characteristic number of positively charged particles called *protons* and negatively charged particles called *electrons*. The proton is about 2000 times as heavy as the electron.
2. The atoms of each element contain the same number of protons as electrons, and therefore are electrically neutral.
3. A dense central portion, called the *nucleus*, contains the protons.

   Also located in the nucleus are particles called *neutrons*. These particles are electrically neutral and are thought by some to be a union of an electron and proton; their weight is essentially that of the proton.
4. Around this nucleus swarm the electrons in a rather definite pattern. The pathways taken by the electrons in their rotation about the nucleus are called *orbits*, or *rings*. The number of rings, or orbits, about the nucleus depends upon the number of electrons.
5. As will be shown, the entire behaviour of the atom is dictated by the number and spatial arrangement of these planetary electrons.

ATOMIC NUMBER

The atomic number of an element is equal to the number of planetary electrons about the nucleus of each of its atoms. Every element has its own characteristic number.

ATOMIC WEIGHT

The atomic weight of an element is equal to the sum of the neutrons and protons in the nucleus of each of its atoms. A glance at the table of elements, however, will disclose that most elements do not have an *even* atomic weight. This is because an element is usually constituted by two or three varieties of atoms which are alike in every respect save the *number of neutrons* in their nuclei. Thus, the atomic weight of chlorine(35)represents the statistical average of its two types of atoms. One type, with 17 protons and *18 neutrons,* has an atomic weight of 35 (constitutes 75 per cent of chlorine), while the other, with 17 protons and *20 neutrons,* has an atomic weight of 37 (constitutes 25 per cent of chlorine). Atoms of the same element which differ *only* in respect to their atomic weights are called *isotopes.* Thus, chlorine has 2 isotopes.

VALENCY

A complete and thorough understanding of valency is essential to the understanding of chemistry. The student, therefore, should devote special attention to the following discussion. Before we define valency, let us see if we can anticipate its meaning. It was stated under the electron theory that an atom's planetary electrons rotate about its nucleus in a rather definite pattern. Substantial evidence indicates that *the orbit, or ring, nearest the nucleus (called the first orbit or, more technically, the k orbit) can contain no more than 2 electrons, and the second and third orbits (I and m respectively) no more than 8 electrons each.*

The maximal number of electrons permitted in each of the subsequent orbits (for example, n, o, p, etc.) may be 8, 18 or 32. Since many of the common and important elements have an atomic number which does not extend beyond the third, or m, orbit, we shall confine our discussion to the 2-8-8 distribution.

The clement *argon* has an atomic number of 18. According to what was said above, the electronic pattern about the nucleus of an atom of argon is thus: 2 electrons are in the first orbit; 8 electrons are in the second orbit; and 8 in the third. The electronic configuration for any element with an atomic number less than that of argon (18) may be drawn by simply allotting the maximal number to the respective orbits. If this is done for all elements from hydrogen (atomic number of 1) to argon (atomic number of 18), *only three elements will have an outer orbit with the maximal number of electrons: helium, neon and argon.*

The outer orbit of helium (which is really its first and only orbit) has the maximal number of 2; the outer orbit of neon (which is really its second orbit) has the maximal number of 8; and the outer orbit of argon (which is really its third orbit) has the maximal number of 8.

Now let us consider the ability of one element to react with some other element. Of the elements with atomic numbers from 1 through 18, only 3 are not capable of reacting with other elements *helium, neon and argon* (these elements are said to be *inert).* It is a most interesting fact that each of these elements are constituted by atoms whose *outer orbits contain the maximal number of electrons.*

In other words, all elements which are chemically active have outer *orbits which do not have the maximal number of electrons.* The chemist has theorized, therefore, that *elements react with one another in order to obtain an outer ring with the maximal number.* The careful reader will readily perceive the tremendous implication of this statement.

The principal implication is this: If elements react for the purpose of obtaining a maximal number of electrons for their atoms (evidently they are "happiest" in this state), we have an excellent explanation of why according to Dalton's Atomic Theory of elements always react in a definite way. For example, sodium has an outer orbit containing 1 electron; chlorine has an outer orbit containing 7 electrons. If the single electron of the sodium atom is taken away and given to the chlorine atom, both atoms will be left with an outer orbit with a maximal number.

In other words, 1 atom of sodium *always* reacts with 1 atom of chlorine to form the compound sodium chloride. If chlorine reacts with aluminum, one atom of aluminum *always* combines (reacts) with 3 atoms of chlorine. This is so because the aluminum atom has 3 electrons in its outer orbit, and must lose these 3 in order to "expose" its second ring which has the maximal number of 8. Therefore, 3 atoms of chlorine are called into play because 1 atom of chlorine can accommodate or receive only one electron. Sodium, aluminum and chlorine are examples of elements which react and thus satisfy their desire for the maximal number of electrons in their outer orbits (usually 8) by a *gain* or *loss* of electrons. Most elements, however, accomplish the same thing by a *mutual sharing* of electrons.

A simple example of mutual sharing is afforded by the reaction between the atoms of hydrogen and oxygen to form water. 2 atoms of hydrogen and 1 atom of oxygen satisfy themselves by coming together and sharing 2 pairs of electrons. In this way oxygen has obtained a maximal number of 8 (8 is the maximal number of the second orbit), and hydrogen a maximal number of 2 (2 is the maximal number of the first orbit). At this time we are prepared to receive the definition of valency.

Valency is defined as the *number of electrons gained or lost by an atom;* or as the *number of pairs of electrons shared by an atom.* Thus, in the above examples, sodium has a valency of 1 (loses 1 electron); chlorine has a valency of 1 (each atom gains an electron); aluminum has a valency of 3 (each atom loses 3 electrons); oxygen has a valency of 2 (each atom shares 2 pairs of electrons); and hydrogen has a valency of 1 (each atom shares 1 pair of electrons). It should be perfectly clear at this point that the valency of an element is its most important chemical property. By knowing an element's valency we can better understand and predict its chemical behaviour. The principal application of valency will be presented in the following chapter. For the present, however, we need to refine our thoughts concerning this important term.

ELECTROVALCNCY

Valency has been defined as the number of electrons gained or lost by an atom, or the number of pairs of electrons shared by an atom. Elements which gain or lose electrons are said to possess an *electrovalency;* those that share electrons, a *covalency.* It should be pointed out at this time, however, that both types of valency are *numerically* equivalent. Thus, when an element with a valency of 2 (whether covalent or electrovalent) reacts with an element whose valency is 1 (whether covalent or electrovalent), they react in the proportion of 1 atom of the former to 2 atoms of the latter. Since this is so, why do we concern ourselves with the two types of valency? Moreover, how do we know that there are two types of atomic combinations? The answer to the second question will be fully explained in a subsequent chapter. As to why we are interested in the two types of valency, it is because electrovalent compounds and covalent compounds have certain fundamental differences. One difference that may be stated at this time is the following: Electrovalent compounds when dissolved in water permit the solution to carry an electric current; covalent compounds do not. Other differences will be pointed out as we go along.

When the sodium atom which is neutral loses an electron it takes on a charge of plus 1 (Na^+). The chlorine atom which accepts this electron takes on a charge of minus 1 (Cl^-). Now we have 2 *oppositely* charged particles which obey the law of electricity which states that unlike charges attract, and like charges repel. The Na^+ and Cl^- particles, therefore, are held together by electrostatic forces. X-ray examination of electrovalent compounds thoroughly substantiates the above explanation. A compound such as sodium chloride is visualized as being constituted by a definite pattern of Na^+ and Cl^- particles.

Elements such as sodium which lose electrons during a chemical reaction are said to have a *positive electrovalency*. For sodium this is expressed as Na^+. Elements which gain electrons are said to have a *negative electrovalency*. For chlorine this is expressed as Cl^-. Similarly, the valency of aluminum is written as Al^{+++} and read as a plus 3. Of extreme importance is the fact that all true *metals* possess a *positive electrovalency;* and all true *nonmetals,* a *negative electrovalency.'* From this we can readily see why metals do not chemically react with metals, and why nonmetals do not react with nonmetals. All electrovalent reactions, then, are between metals and nonmetals. The Table of Valencys presents the important electrovalent elements.

RADICALS

A radical is defined as a *combination of elements which ordinarily acts as a chemical unit.* The term is presented here because radicals are extremely common in chemistry, and possess an electro-valency. All common radicals with the exception of ammonium have a negative electrovalency. Radicals may be considered as "elements." The various radicals and their valencys are presented in the Table of Valencys.*(Radicals never exist in the free state.)*

OXIDATION AND REDUCTION

Throughout chemistry the terms oxidation and reduction are frequently employed. Oxidation when applied to valency (electrovalency) means a *loss of electrons;* reduction when applied to valency means a *gain of electrons.* An element is said to be *oxidized* when it loses electrons, and *reduced* when it gains electrons. In the reaction between sodium and chlorine) the *sodium* becomes oxidized and the *chlorine* reduced. Unfortunately, these two terms are also used to describe reactions involving oxygen. In this connection, oxidation is defined as *the union of oxygen with some other element,* and reduction as *the removal of oxygen from some other element.* The distinction between the above definitions should be thoroughly understood.

COVALENCY

As stated above, covalency means a sharing of electrons. When oxygen combines with hydrogen to form water, each atom of oxygen shares *2 pairs* of electrons, and each atom of hydrogen shares *1 pair* of electrons. Therefore, oxygen has a valency of 2, and hydrogen a valency of 1. Carbon, nitrogen, oxygen, phosphorus and sulfur are other common elements which react or combine via covalency. These elements and their valencys are presented in the Table of Valencys.

There are many elements which possess both an electrovalency and covalency. Hydrogen, sulfur and oxygen are common examples. Hydrogen has an electrovalency of 1 and a covalency of 1; oxygen has an electrovalency of 2 and a covalency of 2; and sulfur an electrovalency of 2, and a covalency of 4 or 6.

MULTIPLE VALENCY

Several elements both electrovalent and co-valent have more than one *set* valency number. Important common elements of this type include: *mercury, copper, iron, sulfur and nitrogen.* These elements are composed of atoms which are capable of losing, gaining, or sharing more than one *set* number of electrons. Iron (Fe), for example, sometimes loses 2 electrons and becomes Fe^{++}, and sometimes 3, and becomes Fe^{+++}.

HYDROGEN

Hydrogen holds a most unique position among the elements. It is the simplest of them all. An atom of hydrogen is constituted by only 1 proton and 1 electron. Since in a chemical reaction the hydrogen atom loses this 1 electron, it has a valency of a plus 1 (H^4). For this reason, hydrogen chemically is considered as a metal.

PERIODIC TABLE

In the latter part of the nineteenth century the Russian chemist, Mendeleeff, thought it would be most interesting if he could juggle the elements around in such a manner that some sort of relationship might be shown. To the amazement of the scientific world, and no doubt to himself, Mendeleeff demonstrated that when the elements (omitting hydrogen) were arranged in order of their atomic weights, there was noted a periodic recurrence of similar chemical properties. In general, the "periods" were of the same length 8 elements.

Now, if these periods were placed under one another, all of these similar elements fell into vertical columns. Mendeleeff's table, however, later proved to have certain discrepancies which could not be resolved in light of the dictates of the Electron Theory. Moseley's work in England in the early part of this century resulted in a revised Periodic Table which obviated the difficulties of the old one. Elements which were obviously out of place in the original table were easily manoeuvered into the correct position when Moseley arranged them according to their atomic numbers rather than their atomic weights.

The original *Periodic Law,* then, which stated that the properties of the elements are periodic functions of their atomic weights, now reads: *The properties of the elements are periodic junctions of their atomic numbers.* The secret of the table will be disclosed if we appreciate the fact that the elements are arranged according to their atomic numbers. Elements of the same vertical column, or *group,* are similar because their atomic numbers are such that the outer orbits of their atoms have the same number of electrons and, hence, the same valency. Thus, the elements lithium, sodium and potassium in Group 1 are alike in their chemistry because each has a valency of 1.

The Periodic Law has proven of incalculable value in the development of chemistry. No doubt its greatest gift to the science has been its suggestions of the existence of certain undiscovered elements, which have subsequently been discovered. In the older tables gaps were present, because often throughout the table there were considerable differences in either the atomic weight or atomic number of neighbouring elements. From the physical and chemical properties of the known elements, the "gap elements" were predicted with uncanny accuracy.

From a student's standpoint, the greatest value of the table resides in its simplification and unification of the whole subject of chemistry. If it were "not for the system of classification employed in biology, the biologist would have no grasp of the astronomical number of diversified organisms which roam the land, fly the air and swim the sea. He would not live long enough, regardless of his intellect, to study all the forms of life and draw conclusions. With his marvelous system of classification, however, it is not necessary for the biologist to study all the creatures.

It is sufficient, indeed, for him to place thousands of similar organisms together into a single class. And so it is true in chemistry. By an appreciation of the implications of the Periodic Table the student always has at hand a powerful tool to cope with the huge number of chemical substances which constitute the earth. Specifically, if one is familiar with the chemical and physical properties of a group, he is familiar with the chemical and physical properties of its members. At appropriate places throughout the text the properties of these groups will be disclosed by a thorough presentation of representative elements.

A few points concerning the Periodic Table require clarification. The Group number (0-VII) indicates the typical valency of all the elements in that particular group. Group 0 is so designated because as explained before these elements have a "full" outer orbit, and do not react with other elements; i.e., they are *inert*. Elements of Group I are constituted by atoms whose outer orbits contain 1 electron; hence they possess a valency of plus 1.

Elements of Group VII are constituted by atoms whose outer orbits contain 7 electrons; hence they have a valency of minus 1. It is evident, therefore, that the true metals are located on the left side of the table (Groups I, II and III), and the true nonmetals on the right side (Groups VI, VII). The elements occupying the central portion of the table (since they contain 4 or 5 electrons in the outer orbits) are covalent. Elements in this area are also *amphoteric;* i.e., they resemble metals in some respects, and nonmetals in other respects.

It is to be noted that each Group, with the exception of Group 0 and Group VIII, is divided into two *Families* A and B. The elements of a family are almost *identical* in their chemical behaviour, but differ considerably sometimes from the elements of its neighbouring family even though the elements of both have the same valency. The elements of Group I, for example, display a closer relation to each other than they do to the elements of any other Group in the table. Within the Group, however, a finer classification exists; i.e., lithium, sodium and potassium, of Family A, are very similar chemically, but differ considerably (chemically) from copper, silver and gold of Family B.

It is to be also noted that the lengths of the 7 *Periods* vary. Periods 2, 3, and 7 are the so-called "short" Periods of 7 or 8 elements in length; 4, 5 and 6 are the so-called "long" Periods of 18 or 19 elements in length; and Period 1 contains only 1 element of hydrogen. Suffice it to say that such an arrangement is necessary in order to have similar elements fall into the same Group.

The elements referred to as the *rare-earth metals* (whose atomic numbers range from 57-71) occupy the same position in Group III of the Table. These 15 rare elements are very similar in their chemical properties. They are only of theoretical interest. In addition to the 92 natural elements, there are 9 *artificial elements* which have been produced in various "atomic machines." These elements follow uranium (atomic number, 92) in the Table. They will be discussed in the chapter on "Nuclear Chemistry." At the present time, therefore, there are 101 elements. There is good reason to believe that additional man-made elements are possible.

Chapter 3

The Molecule, Formula, Reaction and Equation

If it were not for his wonderful system of shorthand, the chemist would indeed be seriously handicapped. The rapid growth of chemistry is due in no small part to the highly efficient methods afforded by the formula and equation. The molecule, formula, reaction and equation are intimately related terms.

THE MOLECULE

Although the atom is our true practical unit of chemistry, we must now consider another unit of practical as well as theoretical importance, the molecule. A molecule may be defined as the *smallest particle of a compound that retains the characteristics of the original substance.* For example, if we were to degrade the familiar compound sodium chloride (our common table salt) into smaller and smaller pieces, theoretically we would arrive at a particle, the molecule, which would be constituted by 2 lonely atoms, sodium and chlorine. This is the smallest particle of salt possible, for if it were to be split any further these 2 atoms would be freed from one another, and our compound would then cease to exist.

It has been shown that our common gaseous elements, oxygen, hydrogen, nitrogen, chlorine, and so on, exist in a condition where 2 atoms occur together rather than singly, as is usually the case with elements. Since this is so, we say that these elements are in the molecular condition. A molecule of oxygen, then, would consist of 2 atoms of oxygen.

THE FORMULA

The chemist says that the formula for salt is NaCl; the formula for water, H_2O; the formula for sulfuric acid, H_2SO_4; and so on. For every known compound the chemist has a formula. What is the meaning and significance of this term? The formula has a definite meaning in chemistry: *It tells us the composition of a molecule.* The Law of Definite Proportions told us, and still tells us, that when atoms of elements combine to form molecules of a compound they always do so in a definite proportion by weight. Another way of saying the same thing is that the same number of atoms is always involved; hence, a molecule of a compound always has the same number of atoms. Since this is so, it is possible to assign a definite statement to each compound concerning its composition the *formula.* Therefore, the formula "NaCl" means that in every molecule of our favourite condiment there is 1 atom of sodium and 1 atom of chlorine; the formula "H_2O" means that in every molecule of water on the earth there are 2 atoms of hydrogen and 1 atom of oxygen; and the formula "HLSO/' means that in every lolecule of sulfuric acid there are 2 atoms of hydrogen, 1 atom of sulfur, and 4 atoms of oxygen.

The next point we should consider is how the chemist knows the kind and number of atoms in the molecule of a certain compound. True, he knows that the number is constant, but this does not indicate the formula. However, the chemist can demonstrate the kinds of atoms in the molecule by the methods of *qualitative analysis.* This is the branch of chemistry which is concerned with the diagnostic chemical and physical properties of elements and Compounds, fin future chapters we shall have occasion to study some of the important aspects of this fascinating branch of chemistry.)

The number of atoms is ascertained by the methods of *quantitative analysis.* In essence, a quantitative study is carried out by taking a known weight of the compound under investigation, and decomposing it into its constituent elements. The quantity of each element found is then expressed on a percentage basis. The number of atoms of the element can now be obtained by dividing the percentage of the element by its atomic weight. For example, ii we were to decompose 9 Ibs. of water by some means, such as an electric current, 1 Ib. of hydrogen and 8 Ibs. of oxygen would be liberated. On a percentage basis, this means that water is composed of 90 per cent oxygen and 10 per cent hydrogen.

Now, if we divide these relative weights, i.e., the percentages, by the relative weights of their respective atoms, i.e., their atomic weights, we shall obtain the number of atoms of each element in a molecule of water. Thus, 90 divided by 16, the atomic weight of oxygen, is 5; 10 divided by 1, the atomic weight of hydrogen, is 10; hence, the ratio of hydrogen atoms to oxygen atoms is 2 to 1, and there are, therefore, 2 atoms of hydrogen and 1 atom of oxygen in every molecule of water.

Fortunately it is not necessary for us to carry out the analysis of every compound to ascertain Us formula. This is only necessary when the chemist is dealing with an unknown substance. The formula for any compound may be written if we know its constituent elements. This is possible by virtue of a concept with which we are already familiar in valency. The rule is this: *Valencys of the elements of a compound must always balance.* The first step is to write down the symbols for the elements present, and next to supply the appropriate number or numbers (called *subscripts)* to the symbols in such a way that the numbers multiplied by the respective valencys produce equal products. For example, what is the formula for hydrogen sulfide? Well, we know from the name that hydrogen and sulfur are the constituent elements, so we write down their symbols, H, S.

The valency of hydrogen is 1; the valency of sulfur, 2. In this form the valencys do not balance, so we must supply the appropriate subscript or subscripts. Just a glance at the problem indicates that by taking 2 atoms of hydrogen to 1 atom of sulfur we have the desired solution, for 2 times the valency of hydrogen, 1, is 2; thus, the correct formula is H_2S. Regardless of the complexity of the compound the general rule applies. If we take another look at the illustration depicting the reaction between sodium and chlorine, we shall readily see why the rule is true. The illustration makes it quite clear that when atoms lose electrons, other atoms must be present to accept them, and vice versa. But even more important is the fact that only a set number of atoms can play the game, since the outer orbit of an atom must have a set number of electrons.

In the case of the sodium-chlorine reaction, the sodium lost 1 electron, so only 1 atom of chlorine, which contains 7 electrons in its outer orbit, was called into play. If chlorine were to react with calcium to form calcium chloride, 2 atoms of chlorine are required, for the calcium atom has 2 electrons to give away, i.e., $CaCl_2$.

The rule for writing formulas applies equally to electrovalent and covalent compounds. As we know, there are a few elements which have variable valencys, depending upon the nature of the other elements with which they react. Thus, sulfur may have a valency of 2, 4 or 6. This does not present a problem since a little experience and practice with the element concerned indicates which of its valencys are involved.

When writing the formula for a compound containing a radical (i.e., SO_4, etc.), it is only necessary to remember that the radical acts as a chemical unit, and has a characteristic valency just as the element. Since the NCX, radical has a valency of 1, 3 of them must be present to balance the valency of aluminum which is 3.

THE REACTION

A great deal has been said about the interaction of atoms; i.e., we have concerned ourselves with the theoretical aspects of chemical activity. Now is the time and place for us to consider activity on a larger scale, i.e., the reaction in the laboratory. But let us never lose sight of the fact that what happens in the test tube is merely an amplification of atomic phenomena.

CHEMICAL REACTION

A *chemical reaction* occurs when two or more substances react with one another to produce new substances. The chemical changes mentioned earlier in the text are now seen to be the result of chemical reactions. Thus, the burning of paper is a chemical change because the elements of the paper carbon, hydrogen and oxygen react with the oxygen in the air to form the new substances, carbon dioxide and water.

Most chemical reactions release energy in the form of heat.

EXOTHERMIC REACTIONS

Energy in the form of light and electricity is released by a few reactions. However, even in these cases heat is usually present. A chemical reaction may liberate its energy gradually, as do those reactions within our bodies, or with explosive violence as when we pull the trigger of a gun. There are some reactions that will not occur unless energy is supplied either in the form of heat, light or electricity. Reactions which require heat in order to take place are called *endothermic* reactions.

THE EQUATION

What happens in the test tube, i.e., the chemical reaction, is succinctly described and explained by the *chemical equation.* The following steps present the manner in which an equation is written:

1. Write the symbol or formula for each of the substances participating in the reaction to the left of an arrow.
2. In a similar manner express the products formed on the right side of the arrow.
3. Balance the equation by placing appropriate numbers (called *coefficients)* before the symbols or formulas in such a way that the number of atoms of each element is the same on both sides of the equation.

Let us see how the above steps apply to a specific reaction, e.g., the reaction between zinc and hydrochloric acid to form zinc chloride and hydrogen. The equation is written thus:

1. $Zn + HC1$
2. $Zn + HC1 \rightarrow ZnCl_2 + H_2$
3. $Zn + 2HC1 \rightarrow ZnCl_2 + H_2$

Our complete balanced equation for the reaction between zinc and hydrochloric acid reads: 1 atom of zinc plus 2 molecules of hydrochloric acid yield J molecule of zinc chloride plus 1 molecule of hydrogen.

In order that we may be able to balance correctly most equations in general chemistry the following points are important to remember:

1. The subscripts of the formulas must *never* be tampered with when balancing. Remember what the Law of Definite Proportions says: The composition of a compound is constant.
2. When dealing with our common gaseous elements whose molecules contain 2 atoms (H_2, O_2, N_2, etc.), they should be written as such in the equation, and not as their symbols (H, O, N, etc.).
3. During a chemical change radicals usually remain intact and are, therefore, treated as chemical units. For example, in the reaction between magnesium sulfate ($MgSO_4$) and sodium hydroxide (NaOH), the sulfate radical, SO_4, and the hydroxyl radical, OH, merely change partners, thus:

$$2NaOH \rightarrow Mg(OH)_2 + Na_2SO_4$$

4. Although an equation is not incorrect without them, arrows are used to indicate when a product is a gas and when a product formed is insoluble. An insoluble product formed as a result of a chemical reaction between solutions is called a *precipitate*.

Types of Reactions

In science we are always classifying in order to facilitate the learning process. Although there are countless chemical reactions, all may be grouped into four types. The types of reactions and illustrative equations are as follows:

1. Direct Combination

   Two or more elements or compounds react to form a more complex substance; also called *synthesis*.

$$C + O_2 = CO_2$$

2. Decomposition

   A reaction in which a compound is broken up into its constituent elements or into simpler compounds; also called *analysis*.

$$2H_2O = 2H_2 + O_2$$

3. Single Replacement or Substitution

   A reaction in which a free element replaces a combined element.

$$Zn + CuSO_4 = ZnSO_4 + Cu$$

4. Double Replacement

   A reaction in which the elements of the reacting compounds change partners. This is probably the most common type of chemical reaction.

IONS

Relatively speaking, we have come a long way in our understanding of matter since beginning. There is still, however, one more concept to assimilate before we have an incipient working knowledge of chemistry. Strangely enough, this concept may be expressed in one single, simple and innocuous word, the *ion*. It is fair to say that the chemist would never have mastered the true nature of the chemical reaction without this word, and what it implies and explains. Although "the ion" antedates the Electron Theory, it was not discussed before because we were not prepared to appreciate its meaning and applications. In this chapter we shall discuss the ion, and those basic points of chemistry which are explained "ionically."

THE IONIZATION THEORY

It has been known for a long time that solutions of certain compounds conduct an electric current and that others do not. Figure below shows the apparatus which is employed to tell whether a solution does.

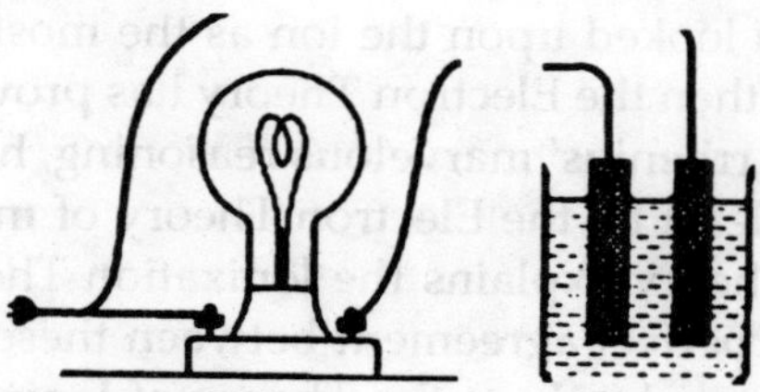

Fig. Apparatus employed to test the ability of a solution to carry a current.

When the bulb lights it shows that a current has passed through the solution. Or does not conduct an electric current. Compounds in solution which allow a current to pass are called *electrolytes;* those which do not are referred to as *nonelectrolytes.* How can we account for this fact? What is the fundamental difference between these two types of compounds? The answer to these questions and many others was presented in the middle of the nineteenth century by the Swedish chemist, Svante Arrhenius, in a monumental piece of work the *lonization Theory.* The lonization Theory, also called the *Theory of Electrolytic Dissociation,* is a rather lengthy piece of work, and well fortified with mathematics. The essence of the theory, though, can be conveniently presented in a manner which is readily intelligible. The following are the cardinal points of the theory:

1. Acids, bases and salts when dissolved in water dissociate *(ionize)* into ions.
2. An ion is an atom or radical with an *electric charge.*
3. *Metallic* ions carry *positive charges,* while *nonmetallic ions* carry *negate c charges.*

4. The solution is electrically *neutral* because the number of positive ions is *equal* to the number of negative ions.
5. When an electric current is passed through the solution, the positive ions, called *cations,* travel toward the negative electrode *(cathode),* while the negative ions, called *anions,* travel toward the positive pole *(anode).*

Arrhenius looked upon the ion as the most fundamental particle. Since then the Electron Theory has proved otherwise, as we know. Arrhenius' marvelous reasoning, however, is not voided in the least by the Electron Theory of matter. Instead, the Electron Theory explains the lonization Theory, and vice versa; there is perfect agreement between these two theories. Let us look more closely at the Theory of lonization, and see how easy it is to understand what it says in the light of what we already know.

There are four kinds of compounds in chemistry, excluding the compounds of carbon: *acids, bases, salts and oxides.* These four kinds of compounds will be adequately treated in the next chapter, but for the moment let us consider acids, bases and salts in relation to the above theory. All acids, bases and salts are electrolytes; this may be easily demonstrated by the conductivity experiment.The reason they conduct a current is offered by the first statement of the lonization Theory which says that these compounds ionize (dissociate) into charged particles called ions. The remaining statements of the theory tell how these ions behave.

But let us go further, and see why these compounds do furnish ions. From our work with valency we know that there are two ways atoms combine *covalency* and *electrovalency.* Since we are looking for charged particles, i.e., ions, we need not bother with covalent compounds, for we said that in this type of combination atoms share electrons; thus, it is the electrovalent compound which holds the secret. Once again our old friend sodium chloride (a salt) has been chosen.

When a molecule of NaCl comes in contact with water, the electrostatic forces which hold the sodium atom to the chlorine atom are broken, and the atoms are freed from one another. But, it is to be noted that after they part they are no longer atoms, *they are ions;* the sodium has one less electron, and the chlorine one more electron than their respective atoms. As a result, the sodium ion has a positive charge of 1 (Na^+), and the chlorine ion a minus charge of 1 (Cl^-). All electrolytes behave in this manner, i.e., all acids, salts and bases.

Let us now define acids, bases and salts on a basis of ionization. *An acid is a compound which in solution furnishes hydrogen ions* (H+); *a base, hydroxyl ions* (OH^-)*; and a salt, metallic and nonmetallic ions e.g., sodium chloride.*

ELECTROLYSIS

Electrolysis is defined as the *decomposition of a compounds by means of an electric current.* It may correctly be inferred that only electrovalent compounds electrolytes are able to be decomposed by this means. Figure illustrates the electrolysis of copper chloride.

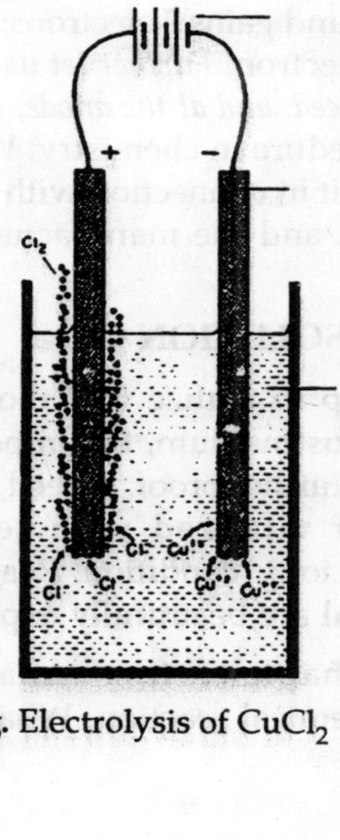

Fig. Electrolysis of $CuCl_2$

When current is applied, the copper ions migrate toward the cathode, and the chlorine ions (Cl^-) toward the anode. At the anode the chlorine ion gives up 1 electron, and becomes an atom. The electrons thus taken by the anode travel through the wire (a current, by the way, is defined as a flow of electrons), and accumulate at the cathode. At the cathode the copper ion takes on 2 electrons, and becomes an atom of copper.

The electrolysis of CuCl. is an example of decomposition, and the equation may be written thus:

$$CuCl_2 \rightarrow Cu + Cl_2$$

Unless ionic equations are specified, the regular equation is satisfactory. The products of the reaction, i.e., copper and chlorine, are formed at the cathode and anode respectively. When this experiment is carried out in the laboratory, bubbles of chlorine gas are seen rising from the anode; the cathode becomes coated with copper.

Electrolysis tells and explains many chemical facts. We should especially note that the number of electrons which the ions gain or lose is equal to their valency. For example, copper has a valency of 2, and gains 2 electrons; chlorine has a valency of 1, and loses 1 electron. *Further let us note that at the cathode an ion becomes reduced, and at the anode, oxidized.* Electrolysis is an important procedure in chemistry. We shall have occasion later on to discuss it in connection with the electroplating and refining of metals, and the manufacture of many important substances.

REACTIONS IN SOLUTION

When we stop to realize that most chemical reactions occur in an aqueous medium, the importance of ions would seem to require no further proof. Indeed, protoplasm is nothing more than water vitalized with certain elements and compounds in the ionic condition. Reactions in solution and reactions in general are beautifully explained via the ion.

In the last chapter a few remarks were presented concerning the chemical reaction. What was not said at that time is implied by this question: How do we know if chemical X will react with chemical Y?

This is an extremely important question because we and the chemist are ultimately interested in getting reactions "to go." If chemical AB and chemical CD are dissolved in water, four kinds of ions will be present in solution: A, B, C and D. Now, these ions are free to move about the solution, and there is every possibility for ion A^+ to come in contact with ion D~ to form AD; and for ion C^+ and ion B~ to form BC. It is to be noted, however, that our "products," AD and BC, ionize as soon as they are formed to A^+, B, C^+ and D, and there is, therefore, every possibility that these ions may combine and form AB and CD, our starting chemicals! If, on the other hand, the products AD and BC were in some way not able to ionize, the starting chemical compounds would not be re-formed.

There are three circumstances under which the products will not react to re-form the original compounds:

1. When one of the products formed is insoluble.
2. When one of the products formed is a gas.
3. When water is formed.

When a product is insoluble or a gas, it is taken from the "field of action," i.e., the solution; when water is formed, it ionizes to such a low extent that the reaction cannot go in the reverse. When the products of a reaction are unable to react to form the original compounds, the reaction is said to be *irreversible;* when they do react to form the original products, the reaction is said to be *reversible.* From a practical point of view, the reversible reaction is no reaction at all nothing happens.

It is the irreversible type which allows us to form new substances. Let us illustrate the reversible and irreversible reactions with specific examples.

Reversible Reactions

If we add a solution of sodium chloride (NaCl) to a solution of potassium nitrate (KNO_3,) the following equation may be written:

$$NaCl + KNO_3 \rightarrow NaNO_3 + KC1$$

Now, the products $NaNO_3$ and KC1 are soluble compounds and ionize readily. Soon, therefore, after the solutions have been mixed, the products are reacting to form the original substances as fast as they are being formed themselves. Actually, it is a dynamic state with reactions taking place in both directions. This reaction, then, and all reversible reactions are written thus:

$$NaCl + KNO_3 \rightarrow NaNO_3 + KC1$$

More scientifically, we say that in a reversible reaction a state of *dynamic equilibrium* exists; i.e., the speed of the forward reaction is equal to the speed of the reverse reaction. Later on we shall have several occasions to utilize the concept of reversibility or equilibrium.

Irreversible Reactions

Let us now add a solution of sodium chloride (NaCl) to a solution of silver nitrate ($AgNO_3$). When these two colourless, clear solutions are mixed, it is instantly perceivable that a chemical reaction has taken place a beautiful white precipitate forms. The reaction is written thus:

$$AgNO_3 + NaCl \rightarrow NaNO_3 + AgCl$$

The white precipitate is silver chloride, and the reaction is irreversible because this compound, being insoluble, is not able to ionize and react with the sodium nitrate ($NaNO_3$). Note that in an irreversible reaction, only the forward arrow is used.

Now let us perform another experiment. If we add zinc (Zn) to a solution of hydrochloric acid (HCL) a reaction takes place instantly: a powerful effervescence ensues; hydrogen gas is liberated. The reaction is as follows:

$$Zn + 2HC1 \rightarrow ZnCl_2 + H_2$$

The reaction is an irreversible one because a gas is formed. At the end of the reaction, only zinc chloride ($ZnCl_2$) remains.

For our last example and type of irreversible reaction, let us add a solution of sodium hydroxide (NaOH) to a solution of hydrochloric acid (HC1). A reaction is evident because a considerable amount of heat is liberated. Also, if we evaporate the reacting mixture, only salt remains. The reaction is represented thus:

$$NaOH + HC1 = NaCl + H_2O$$

This reaction is irreversible because a nonionizable compound is formed.

Let us remember, then, that a reaction goes to completion i.e., all the original substances are consumed to form new products when one of the products formed is a *gas, precipitate or water.* It is just as important, however, to appreciate and remember why this is true.

RATE OF REACTION

By now we should have a fair idea as to what a chemical reaction really is. At this point a few words must be added concerning those outside forces which affect the rate of a reaction. By the rate of a reaction we mean the speed at which a reaction takes place. For example, when powder explodes within a shell, its rate of reaction is tremendously greater than the rusting of iron. Below are presented factors which affect the rate of chemical activity.

Nature of Reacting Substances

Obviously, different substances react at different rates. One point in particular is always true: The speed of reaction between electrovalent compounds greatly exceeds the speed of reaction between covalent compounds. In other words, ions react with one another faster than atoms. This explains why substances in solution usually react at once, as opposed to the slow reactions in the dry state. As a matter of fact, reactions between electrovalent compounds in solution need little or no help from the outside; it is the covalent compound which must be prodded. In the dry state, however, even electrovalent compounds require some energy to react, for in this condition the ions are not as free as in solution.

Temperature

A rise in temperature greatly increases the rate of a reaction. Roughly speaking, a rise of 10° C. about doubles the rate. Therefore, a reaction proceeding at a certain rate at 0° C. will go twice as fast at 10° C., and four times as fast at 20° C. When the temperature is lowered, chemical activity is decreased. Refrigeration is nothing more than a method to retard chemical reactions within foods.

The reason why an increase in temperature speeds a reaction is easy to understand. When substances are heated their molecules become more energetic; hence, they collide with one another more frequently, and strike with greater force. In this manner, more molecules of the reacting substances are given an opportunity to react. As a matter of fact, every reaction of covalent substances has a "critical temperature," below which the molecules of the reacting substances do not have sufficient frequency of collision or impact to react with one another.

Catalysis

Catalysis is a very frequent term in chemistry. We shall employ this word many times before the end of the text. A catalytic agent is defined as an agent which *alters a chemical reaction without itself being changed.* There are many substances which react with difficulty, or not at all, even at high temperatures and under other favourable conditions. If a catalytic agent is added, the reaction will go along smoothly, and usually at a much lower temperature.

Another amazing thing about catalysts is that they are effective in very small quantities; in many cases just a *trace* will enable a chemical reaction to take place between several pounds of reacting substances. Catalysts are also of diverse nature; some are simple compounds (usually metals or oxides), and many are complex carbon compounds of natural origin.

Catalysis, then, is certainly to be considered as a factor which affects the rate or speed of a reaction. The chemist employs catalysts frequently in the laboratory and industry to make substances react, and occasionally to retard chemical activity. The preparation of oxygen will serve to illustrate the importance of catalysis. Oxygen may be prepared by strongly heating potassium chlorate ($KClO_3$) in a test tube. The reaction does not proceed easily, however, and the yield of oxygen is low. On the other hand, if we thoroughly mix a little manganese dioxide (MnO_2) with the potassium chlorate, and then apply heat, the reaction takes place readily and the yield of oxygen is good.

Catalysis is so important that the chemist considers it a subject by itself. Just how a catalyst performs its duties is at the present a moot question. Since a catalytic agent is highly specific, there is perhaps something about its physical and chemical properties which bring the reacting substances into closer molecular and atomic proximity. There is evidence that a catalyst may in some reactions act as an intermediate. In any case, the catalyst remains unchanged at the end of the reaction. This is most fortunate, because chemistry, whether it is concerned with the laboratory, industry, or the tissues of our bodies, strives for economy.

Concentration

The effect of concentration upon a given reaction is succinctly stated by the *Law of Mass Action.* This law states that the *speed of a chemical reaction is directly proportional to the concentration of the reacting substances.* The explanation of this basic and important principle is intuitive. Hypothetically, let us say that substance A reacts with substance B to form C. Further, let us suppose that 1 atom of A and 1 atom of B are placed in a beaker of water. In another beaker let us place 2 atoms of A and 1 atom of B. The question is this: In which beaker are the atoms more likely to meet and react to form C? The answer is almost too obvious: the second beaker. Concentration, therefore, determines the likelihood of atomic interaction or speed of reaction.

OXIDES, ACIDS, BASES AND SALTS

The number of compounds known to the chemist is close to a million, with the list growing rapidly as new synthetic substances are added. As pointed out before, the chemist must employ measures of classification analogous to those of the biologist in order to cope with this bewildering figure. Chemistry is divided into two great areas of study, *organic* and *inorganic.* Inorganic chemistry is the study of all the elements other than carbon; the study of carbon and its compounds is relegated to *organic chemistry.*

The reason why carbon is considered as a separate area of study is because of its strange property of combining with itself. This property is responsible for the huge number of carbon compounds. There are four classes of inorganic compounds: *oxides, acids, bases* and *salts.* Once we have a working knowledge of each class, our study of chemistry will be in clear focus. It is the purpose of this chapter to tell us the principal points of interest of the oxide, acid, base and salt.

OXIDES

An oxide is a compound containing oxygen with another element. Oxygen combines with metals to form metallic oxides, and nonmetals to form nonmetallic oxides. The following are some important oxides: water (H_2O); carbon dioxide (CO_2); carbon monoxide (CO); calcium oxide (quicklime) (CaO); and iron oxide (rust) (Fe_2O_3). Oxides are prepared by burning or heating an element in air.

Acid Anhydride

An acid anhydride is a nonmetallic oxide which reacts with water to form an acid. There is an acid anhydride for most inorganic acids. For example, sulfur trioxide is the acid anhydride of sulfuric acid:

$$S0_3 + H_20 = H_2S0_4\text{, sulfuric acid}$$

Basic Anhydride

A basic anhydride is a metallic oxide which reacts with water to form a base or hydroxide. Calcium oxide is thus the basic anhydride of calcium hydroxide:

$$CaO + H_2O \rightarrow Ca(OH)_2\text{, calcium hydroxide}$$

ACIDS

An acid is best defined as a compound which furnishes *hydrogen ions.* If we remember that an acid is an acid because of the hydrogen ion we will never be confused; it is the FT ion which endows the acid with its characteristic chemical properties. The hydrogen of acids is combined with either one or more nonmetallic elements or a radical. Properties of acids are as follows.

Taste

All acids have a sour taste. The sour taste of lemons, grapes and rhubarb is due to the acid they contain.

Action on Indicators

Certain dyes, called *indicators,* undergo a colour change when exposed to an acid medium. The most popular indicator is litmus. Litmus turns from *blue* to *red* in the presence of acid. Another indicator commonly employed is *methyl orange.* Methyl orange turns from *orange* to *red* in the presence of acid.

Action on Metals

Most metals (copper, mercury, silver, platinum and gold are exceptions) react with acids to form hydrogen and a salt:

$$Zn + H_2SO_4 = ZnSO_4 + H_2$$

In this reaction zinc displaces the hydrogen from sulfuric acid, and forms the salt, zinc sulfate.

Action with Bases

Acids react with bases (metallic hydroxides) to form salts and water. This reaction, called *neutralization,* will be more fully appreciated later on in the chapter. The following is illustrative:

$$\underset{\text{base}}{NaOH} + \underset{\text{acid}}{HCl} \rightarrow \underset{\text{salt}}{NaCl} + \underset{\text{water}}{H_2O}$$

STRONG AND WEAK ACIDS

It is important to understand that acids differ in the degree to which they exhibit the above properties. Acids are, therefore, said to be *strong* or *weak.* Our conductivity experiment is the key to why an acid is strong or weak: *Strong acids are good electrolytes; weak acids are poor electrolytes.* This means that strong acids are highly ionized, and furnish a high concentration of H^+ ions, while the weak acids are slightly ionized, and produce few H^+ ions. The ionic equations for hydrochloric acid (HC1), a strong acid, and acetic acid (CH_3COOH), a weak acid, are written to indicate the degree of ionization:

$$HC1 = H^+ + Cl^- \text{ (highly ionized)}$$

Preparation

The following are the two common methods used in preparing acids:

1. Acid Anhydride plus Water

$$S0_3 + H_20 = H_2S0_4$$

2. Sulfuric Acid plus a Salt

This is the method most applicable to the laboratory. In this procedure, a salt of the desired acid is heated with sulfuric acid. The preparation of nitric acid is illustrative:

$$H_2S0_4 + 2NaNO_3 = 2HNO_3 + Na_2SO_4$$

Nomenclature

A *binary* acid is composed of only two elements, i.e., *hydrogen and some nonmetaL* Binary acids are named by using the stem of the nonmetal with the prefix *hydro-* and the suffix *-ic.* Thus, the acid consisting of hydrogen and chlorine is called *hydrochloric acid;* that consisting of hydrogen and bromine, *hydrobromic;* that consisting of hydrogen and iodine, *hydriodic.*

A *ternary* acid is composed of three elements: *hydrogen, a nonmetal, and usually oxygen.* Three such elements are usually able to combine to produce several acids, i.e., a series. If only two acids are formed, the naming is simple: We use the suffix -/c for the acid containing the more oxygen, and *-ous* for the other. In cases where more than two acids are produced, we must supplement our suffixes with suitable prefixes. For the acid containing the most oxygen, we use the prefix *per-*; for the acid containing the least oxygen, we introduce the prefix *hypo-*.

Let us not be alarmed by the above "chemical grammar." The rules were intended to show that the chemist employs a definite system in naming his compounds. With practice we shall easily add these rules to our working knowledge of chemistry. Moreover, the number of important inorganic acids are relatively few, and easy to remember. The following acids and their formulas should be memorized:

Acid	Formula
Hydrochloric acid	HCl
Sulfuric acid	H_2SO_4
Nitric acid	HNO_3

BASES

A base is a compound of a metallic element with one or more *hydroxyl* (OH) *groups.* A base is best defined, however, as a compound whose water solution contains hydroxyl ions (OH). Let us remember that a base is a base because of the hydroxyl ion, for it is the OH-ion which endows the base with its characteristic chemical properties. Properties are as follows.

Taste

Bases have a bitter taste.

Action on Indicators

Bases act opposite to acids on indicators. Bases turn litmus from *red to blue,* and methyl orange from *orange to yellow.* Another important indicator for bases is *phenolphthalein.* This indicator is turned *red in a basic medium,* and remains *colourless in an acid medium.*

Action with Acids

Bases neutralize acids, forming salts and water. This reaction is called *neutralization.* The following is illustrative:

According to the modern concept *(the Bronsted Theory)*, a base is defined as *an acceptor of protons.* Certain substances, therefore, other than hydroxides may be technically considered bases. For all practical purposes, however, the above definition is entirely satisfactory.

$$NaOH + HC1 \rightarrow NaCl + H_2O$$

Strong and Weak Bases

The strength of a base depends upon the degree of ionization.

Preparation

There are four simple and common methods of preparing bases:

1. Action of an Active Metal with Water

   When sodium or potassium is dropped in water an explosive reaction takes place, with the formation of the hydroxide and hydrogen:

   $$2Na + 2H_2O = 2NaOH + H_2$$

2. Ammonia and Water

   Ammonia reacts with water to form the important base, ammonium hydroxide:

   $$NH_3 + H_2O = NH_4OH$$

Nomenclature

The name of a base is derived from the name of the element, followed by the word hydroxide. The following important bases and their formulas should be memorized:

Hydroxide	Formula
Sodium hydroxide	$NaOH$
Potassium hydroxide	KOH
Calcium hydroxide	$Ca(OH)_2$

SALTS

Salts are compounds which result from the substitution of a metal for hydrogen in an acid. A salt is also defined as a compound consisting of a metal with a nonmetal or an acid radical. Salts far outnumber acids and base? and it is, therefore, most important that we have a good understanding of their properties, methods of preparation and nomenclature. Salts have a wide variety of characteristics. Those containing a common element, however, usually have similar chemical and physical properties. As a general rule, salts are white *crystalline* solids with a "salty" taste. When in solution, all salts are 100 per cent ionized.

Preparation

Salts are prepared in many ways. The following methods of preparation are suggested because of their fundamental nature:

Neutralization

We have already encountered this term twice before. Neutralization is the reaction between an acid and a base, forming a salt and water. Neutralization is also defined as the union of the hydrogen (H+) and the hydroxyl ion (OH~) to form water. (We shall do well to remember that the reason a neutralization reaction goes to completion is because water does not ionize.) The following are neutralization reactions:

$$2KOH + H_2SO_4 \rightarrow K_2SO_4 + 2H_2O$$

$$Ca(OH)_2 + 2HNO_3 \rightarrow Ca(NO_3)_2 + 2H_2O$$

Action of an Acid on a Metal

$$Zn + 2HC1 \rightarrow ZnCl_2 + H_2$$

Action of an Acid on a Metallic Oxide

$$CaO + 2HC1 = CaCl_2 + H_2O$$

Direct Union of Metallic and Nonmetallic Eleiulnts

$$Fe + S= FeS$$

Nomenclature

No difficulty will be experienced in the naming of salts if we recall the nomenclature of the acids. Salts derived from a ternary acid are named by stating the metal and following it with a word having a characteristic suffix. If the name of the acid ends in -/c, the suffix for its salt is *-ate;* if the acid ends in

-ous, the suffix is *-he.* If the name of the ternary acid has a prefix, such as *hypo-* or *per-,* that prefix is made the prefix of the salt. Salts of binary acids are easy to name: the name of the metal is followed by a word having the suffix *-ide.* The table below illustrates the method of naming the sodium salts of the chlorine acids.

TYPES OF SALTS

Acid Salts

Let us consider the neutralization reaction between NaOH and H_2SO_4. In this reaction there are two possibilities:

$$NaOH + H_2SO_4 \rightarrow NaHSO_4 + H_2O \text{ (cold solutions)}$$

Or

$$2NaOH + H_2SO_4 \rightarrow Na_2SO_4 + 2H_2O \text{ (warm solutions)}$$

Thus, acids with more than one replaceable hydrogen atom may form more than one salt. Salts of *poly basic acids* (acids containing more than 1 atom of hydrogen) containing 1 or more atoms of hydrogen are called *acid salts.* Many times acid salts are improperly named. We will experience no confusion in naming an acid salt if we employ the proper prefixes. Thus, the sodium salts of phosphoric acid may be called *trisodium phosphate,* Na_3PO_4; *disodium hydrogen phosphate,* Na_2HPO_4; and *monosodiwn hydrogen sulfate,* NaH_2PO_4. It is common practice to use the prefix, bi-, to name an acid salt of a *dibasic* acid, e.g., H_2CO_3 and H_2SO_4. Thus, Na_2SO_4 is called *sodium sulfate,* and its acid salt, $NaHSO_4$, *sodium bisulfate.*

Normal Salts

A normal salt is a salt which does not contain hydrogen. Thus, in the above illustrations Na_3PO_t and Na_2SO_4 are normal salts.

Double Salts

A double salt is a compound containing two metals combined with one acid radical. Our common *alum,* $KA1(SO_4)_2$, is a good example. This compound is called *potassium aluminum sulfate.* A large number of our mineral substances are double salts.

Hydrolysis

Let us prepare a solution of sodium chloride (NaCl), a solution of aluminum chloride ($A1C1_3$) and a solution of sodium carbonate (Na_2CO_3). If these solutions are now tested with litmus, we shall discover that only the sodium chloride "behaves like a salt," i.e., reacts neutral. The solution of aluminum chloride is acidic, and the sodium carbonate solution is basic in reaction. Since an acid is associated with the H^+ ion, and a base with the OH^- ion, how do we account for the results? The answer to this question is *hydrolysis.*

Hydrolysis is defined as the *reaction between a salt and water producing an acid and a base.* If the acid and base produced are both strong, the solution is neutral; if the acid is stronger than the base, the solution is acidic; and if the base is stronger than the acid, the solution is basic. This explanation is entirely satisfactory, but it will help us to remember and appreciate hydrolysis better if we study the mechanism of the reaction.

It was stated before that water does not appreciably ionize. However, the few ions which are formed are sufficient to react with a salt. Since water ionizes into H^+ ions and OH ions, let us write its formula as HOH. The following reversible reaction takes place in solution:

$$\text{NaCl + HOH = NaOH + HC1}$$

Note that this reaction is reversible because the products are neither insoluble, gaseous nor nonionized. In other words, if the water were to be evaporated from the solution, sodium chloride would remain. Now, both the base (NaOH) and the acid (HC1) are strong electrolytes: i.e., they are highly ionized. The overall picture, then, is that there are as many H ions as OH" ions; hence, the solution is neutral. The following reaction is more complete:

(strong) (strong)
NaCI + HOH = NaOH + HCl

The following is a hydrolysis reaction

(weak) (strong)
$A1C1_3$ + 3HOH = $A1(OH)_3$ + 3HC1

The products, aluminum hydroxide and hydrochloric acid, do not ionize to the same extent. Since HC1 is a strong acid, it furnishes more ions than does the weak base, aluminum hydroxide; thus, the H^+ ions outnumber the OH^- ions, and our solution is acid in reaction.

For the hydrolysis of sodium carbonate solution we have:

$$\underset{\text{(strong)}}{Na_2CO_3} + \underset{\text{(weak)}}{2HOH} = 2NaOH + H_2CO_3$$

Hydrolysis is of great practical concern in and out of the laboratory. Anyone who employs chemistry in any form usually must know if a solution is acidic, basic or neutral. From a theoretical point of view, an understanding of hydrolysis is essential.

HYDROGEN ION CONCENTRATION (PH)

The aphorism that big things come in little packages is thoroughly substantiated by the expression *pH*. *pH* is of paramount interest not only to the chemist, but also to anyone who employs the disciplines of chemistry in any form. The value of *pH* arises from the fact that this term is an *exact* and easy way to express the "reaction" of a solution, i.e., the degree to which it is *acidic or basic.*

There is scarcely a substance which is immune from the influence of a solution with which it comes in contact. Moreover, many substances are inhibited or destroyed by extremely slight changes in acidity or basicity. Of these substances, protoplasm or life itself heads the list; death results when the point of optimum performance of the blood shifts ever so slightly in relation to its acid or base content.

In a liter of pure water the hydrogen ions (H") weigh 0.0000001 gram. *Since water is neutral,* there are as many hydroxyl ions (OH") as hydrogen ions and, therefore, the concentration of hydroxyl ions in a liter of water is the same, or 0.0000001 gram per liter. *In any solution which is not neutral, therefore, the concentrations of the hydrogen ions and the hydroxyl ions will not be equal. A neutral solution has a pH of 7 (pH 7) the number 7 standing for the number of places to the right of the decimal point in 0.0000001.* A solution with an H- concentration of 0.00000001 will have a *pH* 8. Thus, the higher the *pH* the less the acidity or the greater the basicity. The *pH* range extends from 0 (pure acid) to 14 (pure base). A *pH* of 7.5 would indicate that a solution is very slightly alkaline; a *pH* of 6.9 would be on the "acid side of neutral." The *pH* of the blood is between 7.35 and 7.4; a slight change (±:0.5) either way is incompatible with life.

Since the *pH* of a solution is one of its most important characteristics, several methods have been devised to measure it, the most accurate method involving the use of special "*pH* meters." Commonly, however, dye indicators are employed, e.g., litmus, methyl orange and phenol-phthalein. Because these substances do not change colour at an *exact* pH, they are employed only in those cases where the exact *pH* is not required. Litmus, for example, is of value only in so far as it can tell whether a solution is "fairly acid" or "fairly basic." It is possible to buy "pH paper" which affords some degree of accuracy. Weakly acidic or basic solutions which do not cause litmus paper to change may be detected by "pH paper." The test is carried out by wetting the paper, and then matching the colour against a set of standards.

Chapter 4

The Nonmetals

In this chapter we shall study the important nonmetals and their compounds; in the following chapter, the metals and their compounds. We should closely correlate this study with the position of the elements in the Periodic Table. As pointed out before, the Periodic Table lessens our work in chemistry by affording us an easy and applicable system of classification. Briefly, let us review the general arrangement and structure of the table. The vertical columns are called *groups* and the horizontal divisions, *periods.*

Elements of the same group are similar in their chemistry; therefore, if we are familiar with the chemistry of one, we know a great deal about the chemistry of the other elements of the group. This applies beautifully to the first 16 elements, but does not strictly hold throughout the table. For example, the elements copper, silver and gold of Group I bear little relation to lithium, sodium and potassium. Thus, in the same group we may have subgroups; these subgroups are called *families.* In the above example, copper, silver and gold belong to one family (Family B in the table), and lithium, sodium and potassium another (Family A). Each group has its characteristic valency. Since some elements exhibit more than one valency, the valency indicated above each group is the one which is most common to the elements of that group.

It is to be recalled also that elements of Group I are true metals, and those of Group VII, true nonmetals. As we pass toward the middle of the table from either side, the elements become somewhat ambiguous in nature. Elements of Group VIII are metallic in their physical properties and variable in their chemical properties. The elements of Group 0 are the so-called inert gases; since their outer orbits are filled, they do not react with other elements.

OXYGEN

Oxygen is the most abundant element. It occurs in the free state in the atmosphere at a concentration of about 21 per cent, and in the combined state, e.g., water, quartz, clay and limestone, it constitutes 50 per cent of the earth's crust. The credit for the discovery of oxygen is given to Joseph Priestley, an English clergyman. Priestley prepared the gas in 1774 by heating mercuric oxide (HgO).

PREPARATION

Oxygen may be prepared in several ways. The following two methods are most applicable to the laboratory:

Decomposition of Mercuric Oxide

$$2HgO \rightarrow 2Hg + O_2$$

Decomposition of Potassium Chlorate

This is the most common method. In this method a mixture of potassium chlorate ($KClO_3$) and manganese dioxide (MnO_2) is gently heated. Since the gas is insoluble in water, it is collected by the method

$$2KClO_3 = 2KCl + 3O_2$$

In this reaction manganese dioxide acts as a catalyst.

Commercially, oxygen is prepared by the following two methods:

Electrolysis of Water

When an electric current is passed through water (a little H_2SO_4 is added to make the water conductive), it is decomposed into oxygen and hydrogen. The hydrogen is collected over the cathode, and the oxygen over the anode:

$$2H_2O \rightarrow 2H_2 + O_2$$

Liquid Air

When air is subjected to extremely high pressures and low temperatures, it is converted into a liquid. At room temperature liquid air boils away. Since nitrogen has a lower boiling point than oxygen, it escapes first, leaving pure oxygen.

Physical Properties

Oxygen is a colourless, odorless and tasteless gas. The gas is slightly soluble in water, 100 cc. of water at 0° C. dissolving 4 cc. of the gas at atmospheric pressure. Oxygen is slightly heavier than air.

Chemical Properties

Oxygen combines with most of the elements nonmetals as well as metals to form oxides. This process is called *oxidation*. Let us recall another definition of this term: Oxidation is the loss of electrons. When we think of oxidation another term suggests itself *reduction*. Reduction is the opposite of oxidation; it likewise has two definitions: Reduction is defined as the gain of electrons, or as the loss of oxygen. Oxidation and reduction, then, may or may not involve oxygen.

An agent which brings about the oxidation of another substance is called an *oxidizing agent;* one that reduces another substance is called a *reducing agent*. These agents may or may not involve oxygen. For the moment we are concerned with oxidation and reduction and their agents as they apply to oxygen. The following reactions illustrate oxidation and reduction:

Oxidation

$$4Fe + 3O_2 = 2Fe_2O_3,$$

Reduction

$$CuO + H_2 = Cu + H_2O$$

In these reactions the following points are to be noted:

1. Iron (Fe) is being *oxidized* by oxygen; i e., oxygen is the *oxidizing agent.*

2. Copper oxide is being *reduced* by hydrogen; i.e., hydrogen is the *reducing agent.*
3. Further, note that when one substance is oxidized another substance is reduced, and vice versa. Thus, while the CuO is being reduced, the hydrogen is being oxidized.

Oxidation is said to be *slow* or *rapid.* The rusting of iron and the "burning" of sugar within our tissues are examples of slow oxidation. Rapid oxidation, or *combustion,* is synonymous with burning. In contrast to slow oxidation, rapid oxidation causes light and heat to be emitted. *Spontaneous combustion,* however, testifies to the fact that heat is liberated during slow oxidation. If oily rags are placed in a dark warm room with poor ventilation, the heat which accumulates from the oxidation of the oil will reach ah intensity sufficient for the rags to burn. This brings up another important point of *kindling temperature.* The kindling temperature is defined as the temperature to which a substance must be heated in order to burn. The more combustible a substance is, the lower is its kindling temperature.

Strangely enough, oxygen does not burn, it supports burning. This is not a paradox, however, when we realize that burning, i.e., the liberation of heat and light, is merely the manifestation of a chemical union between oxygen and some other element.

Uses

Oxygen is placed into steel cylinders under pressure. It is employed extensively in medicine in cases of respiratory embarrassment. In industry oxygen is widely used to produce high temperatures for welding and cutting. The oxyhydrogen torch and oxyacetylene torch reach temperatures as high as 4000° C.

OZONE

Several elements, e.g., oxygen, phosphorus, sulfur and carbon, exist in two or more forms. Although their properties are essentially the same, they do have individual characteristics. One form is always more stable and common than the others. These forms of the same element are called *allotropic* forms.

Ozone is written O_3 and is the allotropic modification of oxygen. Ozone is a colourless gas possessing a garlic-like odor; liquid ozone is dark blue. Ozone may be readily prepared by passing a *powerful spark through air or oxygen.* The reaction is as follows:

$$3O_2 = 2O_3$$

Chemically, ozone is a *much more powerful oxidizing agent than oxygen.* For this reason, ozone is often used where an oxidizing agent is indicated. In the form of ozonized air is produced by a silent electric discharge, it has been used to sterilize, bleach and remove objectionable odors.

WATER

Water, the principal compound of oxygen is our most familiar, most abundant, most useful and most interesting chemical compound. Water not only covers most of the earth's surface, but is also the *principal constituent of its plants and animals: Cucumbers and tomatoes are about 95 per cent water; the human body is about 70 per cent water.* Water is without question the marvel of the earth, and perhaps the universe. Water is a marvel to the chemist because of its multitude of useful and unique properties, in conjunction with the fact that it is one of the earth's simplest substances H_2O.

PHYSICAL PROPERTIES

Water is a tasteless and colourless liquid. At standard pressure (1 atmosphere or 14.7 Ibs. per sq. in.), water freezes at 0° C. (32° F.) and boils at 100° C. (212° F.). Water has the peculiar property of expanding when it cools from 4° to 0° C., and while it freezes. At 4° C. water has its greatest density: 1 cc. weighs 1 gm Above or below this temperature 1 cc. weighs less (very slightly of course) than 1 gm.

CHEMICAL PROPERTIES

Although water displays extreme stability when subjected to high temperatures (it may be heated to 1000° C. before any decomposition takes place), it does react with several chemical substances to produce important products.

Action with Metals

Water reacts with the active metals such as potassium, sodium, calcium and iron. In this action part or all of the hydrogen is replaced. Sodium, for example, reacts with water to produce sodium hydroxide and hydrogen, thus:

$$2Na + 2H_20 = 2NaOH + H_2$$

Action with Acid Anhydrides

It was pointed out before that nonmetallic oxides, or acid anhydrides, react with water to produce an acid:

$$S0_3 + H_2O = H_2SO_4$$

Action with Basic Anhydrides

Water reacts with metallic oxides, or basic anhydrides, to produce bases or hydroxides, thus:

$$CaO + H_20 = Ca(OH)_2$$

Hydrolysis

As we already know, salts in solution render the solution acidic, basic or neutral, depending on the products formed in the equilibrium reaction between the salt and water (hydrolysis). A solution of ammonium chloride, for example, is acidic because the hydrochloric acid formed is a stronger acid than the ammonium hydroxide formed is a base:

$$NH_4C1 + HOH = NH_4OH + HC1$$

Electrolysis

Water is decomposed by an electric current into its constituent elements: hydrogen and oxygen.

Water of Crystallization

A characteristic property of most salts is that they possess a crystalline structure. If certain crystalline salts are heated they change to a shapeless or amorphous mass. It will be discovered during the course of heating that water is driven off. This water, called the *water of crystallization,* or *water of hydration,* is an essential constituent in the structure of the crystal. The powder remaining after the water of crystallization has been removed is referred to as the *anhydrous* form of the compound. The term *hydrous* is applied to the crystalline form. Many substances contain water, but in a *mechanical way* and, therefore, in varying amounts.

Many hydrous salts lose water when merely exposed to the air. Such substances are said to be *efflorescent.* Crystals of washing soda on standing in air lose their water of crystallization and crumble to a white powder. Efflorescence, therefore, has an economic as well as a chemical meaning; namely, that when a hydrous salt is purchased, you are purchasing a certain amount of water. Two other terms come to mind when efflorescence is mentioned: *deliquescent* and *hygroscopic.* A hygroscopic substance is one which absorbs moisture from the air. A deliquescent substance is one which absorbs sufficient water from the air to *dissolve itself.* Calcium chloride, for example, is so deliquescent that a crystal liquefies soon after being exposed to the atmosphere.

Purification

In order to make water fit for human consumption it is treated physically and chemically to remove any harmful foreign matter. In addition, water should be clear and palatable.

HYDROGEN PEROXIDE

Oxygen forms two oxides with hydrogen: *hydrogen monoxide* (water); and *hydrogen dioxide* or, more commonly, *hydrogen peroxide.* The latter compound does not obey the law of valency.

PREPARATION

Hydrogen peroxide is prepared by the reaction between dilute sulfuric acid and barium peroxide, thus:

$$H_2SO_4 + BaO_2 \rightarrow H_20_2 + BaSO_4$$

The above reaction goes to completion because barium sulfate is insoluble. Hydrogen peroxide is recovered from the reaction by filtration. Properties. Pure hydrogen peroxide is a clear, syrupy liquid about one and one-half times as heavy as water. Since it holds its "excess" oxygen very loosely, it is a powerful oxidizing agent. In fact, it is so unstable that it readily explodes when exposed to heat or light. Hydrogen peroxide decomposes according to the following equation:

$$H_2O_2 = H_2O + O \text{ (nascent oxygen)}$$

USES

As pointed out before, the nascent state of a gas i.e., the state in which the atoms of a gas are set free, and before they combine to form molecules is highly active chemically. This activity is probably explained by the energy released when the atoms unite, thus:

$$2O \rightarrow O_2 + \text{energy}$$

For this reason, hydrogen peroxide possesses antiseptic and bleaching actions. Its bleaching action, however, is confined to those pigments which are decolourized by oxidation. Hydrogen peroxide is sold in the pharmacy as a 3 per cent solution of the pure compound in water. Obviously, aging and heat will attenuate its action. At best, hydrogen peroxide is only mildly antiseptic. The effervescence produced (escape of oxygen) when hydrogen peroxide is applied to a wound is no doubt more important than its inherent antiseptic action. The effervescence aids in loosening tissue debris, and thereby facilitates cleansing.

SOLUTIONS

The solution is one of the most practical tools in chemistry. A knowledge of the principles of solutions is essential to anyone who employs chemicals. *A true solution is an homogeneous liquid mixture of two or more substances whose particles are of molecular or atomic size.* The *solute* refers to the substance which goes into solution; it may be either a solid, liquid or gas. The *solvent* is the substance (usually water) which dissolves the solute. When two liquids mix in all proportions, thus forming a true solution, they are said to be *miscible,* e.g., alcohol and water. Nonmiscible, or *immiscible,* liquids do not mix, e.g., oil and water. The terms *suspension* and *emulsion* should be differentiated from solution or true solution. A suspension is a mixture of a solid and a liquid in which the particles settle out on standing, e.g., sand in water. An emulsion is a mixture of two or more immiscible liquids which separate slowly, e.g., milk.

Kinds of Solutions. A general and easy way to classify solutions is on a basis of the proportion of solute to solvent. Later on we shall learn more precise ways to express a solution's concentration.

1. *Dilute solution* is one which contains a small amount of solute in relation to the solvent.
2. A *concentrated solution* is one which contains a large amount of solute in relation to the solvent.
3. An *unsaturated solution* is one which is capable of dissolving more solute under the same conditions of temperature and pressure.
4. A *saturated solution* is one which contains all the solute it can normally dissolve at a given temperature and pressure.
5. A *supersaturated solution* is of little practical importance, but of some interest. This type of solution may be easily demonstrated in the laboratory by using a chemical called sodium thiosulfate (hypo). If a saturated solution of hypo (at a given temperature) is heated, additional solute can be dissolved. This is common experience. Now, if this solution is carefully cooled to the original temperature, the hypo which

was dissolved over and above the saturation point remains in solution.

Such a solution is said to be supersaturated. This condition is extremely unstable. If a minute crystal of hypo is added (or if the vessel is shaken) the excess solute will crystallize out, leaving a saturated solution.

PREPARATION OF SOLUTIONS

The speed with which a solid goes into solution is usually increased by *pulverization, agitation* (e.g., stirring) and *heating*. An increase in the temperature of the solvent *decreases* the solubility of gases. An increase in pressure, however, *increases* the solubility of a gas. Soda water, for example, contains dissolved carbon dioxide which is put in solution and kept there by pressure. The effervescence (escape of gas) seen when a bottle of pop is opened is due to the release of pressure. Pressure has very little effect on the solubility of solids. The miscibility of liquids is not significantly altered by temperature or pressure.

COLLIGATIVE PROPERTIES

Solutions have certain properties which are related to- the concentration of the solute. These properties, *freezing point, boiling point* and *osmotic pressure* are known *gative properties* of the solution.

FREEZING POINT

Water, which freezes at 0° C. in the pure state, freezes at a lower temperature when it contains a dissolved substance. This applies to all solutions. *The freezing point of a solution is lower than that of the solvent,* and the degree of depression is proportional to the concentration of the solute.

BOILING POINT

The boiling point of a solution is *higher* than that of its solvent, the degree of elevation being proportional to the concentration of the solute.

OSMOTIC PRESSURE

Before we can appreciate what is meant by a solution's osmotic pressure, we must understand the term *osmosis.* Osmosis may be simply and adequately explained by an apparatus. This apparatus consists of a thistle tube to the lower end of which is attached a parchment membrane. Within the tube is a solution of sugar in water. The parchment serves as a *semi-permeable membrane.* A semipermeable membrane is a membrane which permits solvent molecules to pass (water), but not solute molecules (sugar). Since the concentration of water outside the tube is greater than it is inside, water will diffuse from *outside* to *inside* in an attempt to equalize these concentrations. This "migration" from outside to inside is evidenced by the rising of the water in the glass tube.

The water will stop rising when the hydrostatic pressure is sufficient to prevent further passage of solvent molecules. *Osmosis, then, is the diffusion which proceeds through a semipermeable membrane, separating two solutions, with the principal flow toward the denser solution. Osmotic pressure* refers to the "pull" the denser solution has on the less dense solution. It is important to remember that osmosis occurs whenever two miscible solutions *without regard to the nature of the solutes* are separated by a semipermeable membrane. Living cells have semipermeable membranes, and depend for much of their activity upon osmosis. In our study of biochemistry we shall have many occasions to employ this term.

NITROGEN AND ITS COMPOUNDS

Nitrogen is the most abundant element of the atmosphere: It constitutes about 78 per cent of the air by volume. In the combined state it occurs in a number of compounds, both organic and inorganic. The chief inorganic forms of nitrogen are mineral nitrates, specifically, sodium nitrate (Chile saltpeter). Organically, nitrogen is the characteristic constituent of proteins and, therefore, is present in the protoplasm of all plants and animals. Presently we shall see that nitrogen is just as vital to our existence as oxygen.

PREPARATION

Removal of Oxygen from Air

This is an excellent way to prepare nitrogen in the laboratory. The easiest manner in which we can remove oxygen from the air is to place a piece of phosphorus in an enclosed volume of air. The phosphorus reacts readily with the oxygen to form phosphorus pentoxide, a solid; the gas remaining is almost pure nitrogen:

$$4P + 5O_2 \rightarrow 2P_2O_5$$

Liquid Air

This is the commercial method of preparation. Air is first reduced to a liquid by high pressures and low temperatures. When the pressure on the liquid is released the nitrogen readily escapes. The escaping nitrogen can then be liquefied and placed into steel cylinders under pressure.

PHYSICAL PROPERTIES

Nitrogen is a colourless, tasteless and odorless gas. It is slightly lighter than air, and slightly soluble in water.

CHEMICAL PROPERTIES

Chemically, nitrogen is not a very active substance. Nitrogen does not burn, support combustion, or react readily with other elements. It does under extreme conditions combine with oxygen to form oxides, and with hydrogen to form ammonia.

THE ATMOSPHERE

Something should be said about the gaseous sea within which we live the air. Minute traces of air are probably to be found at a distance as high as 2000 miles above the earth. The bulk of the air, however, lies within a range of five miles above the earth. At sea level air exerts a pressure of 14.7 pounds per square inch. Since the composition of air is not exactly constant, and since its constituents exert their individual characteristics, *air is certainly a mixture and not a compound.* The average composition of dry air is as follows:

Nitrogen 78; Oxygen 21%; Carbon dioxide 04%; Argon %%; Neon; Helium; Krypton; Xenon; Hvdrogen; Ozone

THE NITROGEN CYCLE

In nature, nitrogen is necessary for life. A complex relationship exists between the plant and animal worlds relative to this element. Essentially, this is a reciprocal agreement whereby the plants feed the animals and the animals, in turn, aid in maintaining the plants. The free nitrogen of the air, strangely enough, is not available to plants and must, therefore, be converted into a usable form. The related chemical events, involving the free nitrogen of the air and its compounds in protoplasm and in the earth's soil, constitute the so-called *nitrogen cycle.*

Briefly, the cycle may be described thus: A variety of microorganisms in the soil bring about the *decay* or *putrefaction* of organic matter, i.e., material derived from dead plants and animals and animal wastes. As a result of this microbial action, *ammonia* is liberated. Another group of microorganisms, called *nitrifying bacteria,* convert this ammonia into *soluble nitrate salts,* e.g., sodium nitrate. Plants absorb these nitrates into their roots and utilize them to manufacture proteins for protoplasm. The animal world obtains its proteins by consuming plants. In time plants and animals die and add their nitrogen to the soil in the form of nitrates.

The above situation is somewhat complicated by the fact that another group of bacteria, called *denitrifying bacteria,* act upon nitrates to form nitrogen gas, which escapes to the air. If there were not some mechanism, therefore, whereby this lost nitrogen could be returned to the soil, the earth would become depleted of its nitrates and life. A strange group of bacteria, called *nitrogen-fixing* bacteria, are able to convert, directly, atmospheric nitrogen into bacterial protoplasm. Since these organisms live a symbiotic life on the roots of certain plants (e.g., *alfalfa* and *clover),* they share their proteins or protoplasm with the plant. The return of nitrogen to the soil by the nitrogen-fixing bacteria completes the cycle.

AMMONIA (NH_3)

Ammonia, a gaseous compound of nitrogen and hydrogen, is probably the most important compound of nitrogen certainly the most popular. It is present in minute quantities in the air as a result of the decomposition of plant and animal matter. The characteristic odor of decaying manure is due in no small part to the ammonia being liberated.

Laboratory Preparation

In the laboratory ammonia is conveniently prepared by heating a mixture of ammonium chloride (NH_4C1) and calcium hydroxide ($Ca(OH)_2$). These two compounds react (double decomposition) to form ammonium hydroxide (NH_4OH) which, being unstable, decomposes into ammonia gas and water. The ammonia, being lighter than air, is collected by upward displacement. The reaction is as follows:

$$2NH_4C1 + Ca(OH)_2 = CaCl_2 + 2NH_4OH$$

Or

$$2NH_4C1 + Ca(OH)_2 = CaCl_2 + 2H_2O + 2NH_3$$

Commercial Preparation

Ammonia is prepared commercially in vast quantities chiefly by three processes: *destructive distillation* of soft coal, i.e., heating coal out of contact with air, the *Cyanamide Process*, and the *Haber Process*. The Haber Process heralded an era of synthetic chemistry. From a practical and theoretical point of view, this process is of paramount interest to us. The Haber Process employs nitrogen from liquid air, and hydrogen from the electrolysis of water. These elements are made to combine as follows:

$$N2 + 3H_2 = 2NH_3$$

This is a most interesting situation. It was stated before that we do not receive yields from reversible reactions. However, it was also pointed out that reversible equations may be rendered "productive" by the application of certain principles. As the equation stands, it is true nothing does happen. If we increase the *temperature* and the *pressure,* however, ammonia is formed. The increase in pressure is especially important. In this equation and in all equations involving gases the reaction may be stated in terms of volumes, thus: 1 volume of nitrogen reacts with 3 volumes of hydrogen to produce 2 volumes of ammonia. In other words, the volume of the ammonia is one-half the total volume of the reactants.

This is the reason an increase in pressure favours the formation of ammonia. Perhaps we can understand this effect of pressure more effectively if we recall what happens when a piece of ice is squeezed in the hand: The ice melts in a shorter period of time than if the ice is simply held in the hand. Since the volume of the ice exceeds the volume of the water, an analogy is evident: *Pressure favours the formation of a product whose volume is less than the volume of the original substances.* In a general sort of way this makes sense, because when a "system," having been subjected to pressure, assumes a smaller volume, it is "absorbing" or "undoing" the pressure so as to relieve the stress.

This effect of pressure is a special case of one of the most important generalizations of chemistry and physics, first stated by Le Chatelier in 1887. *Le Chatelier's Principle* is stated as follows: *a stress is applied to a system in equilibrium, the position of the equilibrium will shift in such a direction as to diminish the applied stress.* The Law of Mass Action (p. 34) is another special case of this principle. Thus, by applying Le Chatelier's Principle, many "worthless equilibria" are made important industrial processes. Under the favourable conditions of the Haber Process, then, the above equation becomes:

$$N_2 + 3H_2 = 2NH_3$$

Physical Properties

Ammonia is a colourless gas, having a strong pungent odor. It is approximately one-half times as heavy as air. Ammonia liquefies easily. One of the outstanding characteristics of ammonia is its extreme solubility in water: *One volume of water dissolves 700 volumes of ammonia.*

Chemical Properties

Ammonia burns in pure oxygen to form water and nitrogen; it reacts directly with acids forming *ammonium* salts:

$$4NH_3 + 3O_2 = 6H_2O + 2N_2$$

$$NH_3 + HC1 = NH_4C1, \text{ ammonium chloride}$$

Another important property of ammonia is its reaction with water, forming ammonium hydroxide:

$$NH_3 + H_20 = NH_4OH$$

Ammonium hydroxide, as might be inferred from the above reaction, is an unstable chemical. Heat or reduced pressure favours the backward reaction, forming ammonia and water. Ammonium hydroxide is a very weak base.

USES

Huge quantities of ammonia are employed in the manufacture of nitric acid, ammonium hydroxide (household ammonia) and ammonium sulfate. The latter compound is a common constituent of fertilizer.

SULFUR AND ITS COMPOUNDS

Sulfur is as old as history, and has played a vital part in man's conquest of matter. There is scarcely a laboratory or industry which does not employ sulfur or its compounds. Sulfur itself from a chemical standpoint is our most important raw material. The yearly production of sulfur in the United States alone is nearly 4,000,000 tons.

FRASCH PROCESS

Free sulfur occurs in huge deposits in the earth. The chief centers of production in the United States are in Texas and Louisiana; in Europe the chief centers are in Italy and Sicily. Although nature was kind enough to supply us with sulfur in the free state thus obviating costly processes we still had to solve the very difficult problem of mining it. We can easily understand the difficulty when we realize that sulfur in this country occurs under about 700 ft. of quicksand, thus making ordinary methods of mining impossible. Prior to 1900, our sulfur was imported because no one had solved the problem.

Then in 1891 Frasch, an American engineer, supplied us with an ingenious answer. The Frasch Process has enabled us to raise our production from almost zero tonnage to nearly 4,000,000. Three pipes are sunk to the sulfur level. Water heated under pressure to about 170° C. is forced down the outermost pipe. The sulfur is thereby melted. Hot air is then forced down the innermost pipe, mixing the melted sulfur to make it foamy and light. The mixture is then piped into wooden bins where solid sulfur separates. Sulfur mined in this manner is 99.5 per cent pure!

Not all sulfur is in the free state. Sulfur occurs in the mineral world in many important sulfides and sulfates; in the organic world sulfur is a vital element in the fabric of many plant and animal proteins.

PHYSICAL PROPERTIES

Commercial sulfur is a golden-yellow solid with little taste or odor. It is insoluble in water, but soluble in carbon disulfide and certain other organic solvents. Sulfur melts at 119° C., and boils at 444.6° C. Sulfur exists in *five different allotropic forms,* three of which are solid, and two liquid.

SOLID SULFUR

Of the three solid forms, two are crystalline, and one amorphous (noncrystalline). *Rhombic* sulfur is the most stable form of sulfur and is, therefore, the form which all sulfur assumes at room temperature. Rhombic sulfur is prepared by dissolving a little sulfur in carbon disulfide, and allowing this solution to stand in an evaporating dish; pure and perfect crystals remain. *Prismatic* (also called monoclinic) crystals can readily be obtained by melting some sulfur in a small crucible, allowing it to cool so that it partially solidifies, then pouring off the remaining liquid. The needle-shaped crystals adhering to the bottom and walls belong to the prismatic system. *Amorphous* sulfur is prepared by pouring boiling sulfur into cold water. This form of sulfur is a rubbery, amorphous mass, which returns to the rhombic form upon standing.

LIQUID SULFUR

Just above its melting point sulfur is a thin, light yellow liquid. This is one form of sulfur. As the temperature is raised, this liquid becomes dark, and at about 200° C. becomes viscous. On further heating, the liquid becomes thin again, and boils at 444.6° C. This dark liquid is the second form.

CHEMICAL PROPERTIES

Sulfur has an atomic structure similar to oxygen and, therefore, has chemical properties much like that element. Both elements occur in the same family (Family B, Group VI). Oxygen forms oxides with metals; sulfur forms sulfides:

$$Fe + S \rightarrow FeS$$

$$Ag + S \rightarrow AgS$$

Like oxygen, sulfur vapour supports combustion. For example, hydrogen burns in a vapour of sulfur:

$$H_2 + S \rightarrow H_2S$$

Sulfur reacts with nonmetals as well as with metals. Thus, sulfur reacts with oxygen (i.e., burns) to form sulfur dioxide gas. Sulfur burns with a pale blue flame:

$$S + 0_2 = SO$$

HYDROGEN SULFIDE (H_2S)

Hydrogen sulfide is a highly odoriferous gas formed by decaying organic matter which contains sulfur. The characteristic odor of the water from "sulfur" springs is due to this gas. Hydrogen sulfide is also a constituent of the gases issuing from volcanoes.

Preparation

Hydrogen sulfide is readily prepared in the laboratory by the action of dilute hydrochloric or sulfuric acid on a sulfide:

$$FeS + 2HC1 = FeC12 + H_2S$$

Physical Properties

Hydrogen sulfide has the offensive odor of rotten eggs. It is heavier than air and slightly soluble in water.

CHEMICAL PROPERTIES

Hydrogen sulfide burns readily:

$$2H_2S + 30_2 = 2H_20 + 2S0_2$$

In water it dissolves sparingly to form hydrosulfuric acid (H_2S). The most characteristic property of hydrogen sulfide is its reaction with certain metallic ions to *form insoluble coloured sulfides.* When this gas is bubbled through a solution of copper sulfate, for example, a black precipitate forms immediately. The reaction is one of double decomposition between hydrosulfuric acid (the gas in water forms the acid) and the metallic salt:

$$CuS0_4 + H_2S = H_2SO_4 + CuSS$$

Use

Hydrogen sulfide finds its principal value in qualitative chemical analysis, where it is used to identify those metallic ions with which it forms characteristic sulfides. The following equations are illustrative:

$ZnCl_2, + H_2S = 2HC1 + ZnS,$
white

$CuSO_4 + H_2S = H_2S0_4 + CuS$
black

$2BiCl_3 + 3H_2S = 6HC1 + Bi_2S_3$
brown

$2SbCl3 + 3H_2S = 6HC1 + Sb_2S_3$
orange

Test for a Sulfide. To identify a sulfide we carry out the following test: A little dilute hydrochloric acid is added to the compound under investigation, and then a strip of paper moistened with lead acetate solution is held over the mouth of the test tube. If the paper turns dark, the unknown is a sulfide. As discussed above, the acid reacts with the sulfide to form hydrogen sulfide, and the hydrogen sulfide then reacts with the lead acetate to form the black compound, lead sulfide. The test is illustrated by the following equations:

$$FeS + 2HC1= FeCl_2, + H_2S$$

SULFUR DIOXIDE (SO_2)

Sulfur dioxide is found to a limited extent in volcanic gases. It is also evolved by burning coal, which usually contains small quantities of sulfur, whether free or in the combined form.

Preparation

In the laboratory sulfur dioxide is readily prepared by gently heating a *sulfite or a bisulfite with an acid.* The reaction produces sulfurous acid (H,SO_3), which immediately decomposes into water and sulfur dioxide.

SULFURIC ACID (H_2SO_4)

Sulfuric acid is unquestionably the "king of compounds." Directly or indirectly, it is used in practically every industry. It is an economic fact that of a country's degree of civilization bears a direct proportion to the consumption of this vital chemical. Sulfuric acid is the most typical of the typical acids.

PREPARATION

Most of the sulfuric acid at the present time is made by the *contact process.* This is an important process not only for its industrial worth, but also because the student is supplied with an example of a good (easy to understand) industrial synthetic procedure. Every step in the sequence of events employs a familiar chemical reaction. The industrial method is nothing more than a chemical amplification of this working model. The steps are as follows:

1. Sulfur dioxide is produced by burning sulfur.
2. A mixture of this sulfur dioxide and air (O_2) is cleaned and dried by passing through cotton and sulfuric acid, respectively (sulfuric acid is a dehydrating agent) .

3. Sulfur trioxide is formed when the sulfur dioxide and air are passed over the heated catalyst (platinum).
4. The sulfur trioxide is then absorbed by concentrated sulfuric acid to form fuming acid.
5. The fuming sulfuric acid is then diluted with water to form concentrated sulfuric acid.

PROPERTIES

Concentrated sulfuric acid is a heavy, syrupy and corrosive liquid. It has a boiling point of 338° C., and is thus the least volatile of all ordinary acids. Much heat is evolved when the acid and water are mixed. When preparing the dilute solution, the *acid is added to the water, never the reverse to avoid dangerous spattering*. Even with this procedure the vessel becomes very hot.

The chemical properties of sulfuric acid depend on whether it is dilute or concentrated. The dilute acid is a typical acid; it reacts with metals to liberate hydrogen, and neutralizes hydroxides. Concentrated .sulfuric acid is a powerful *oxidizing and dehydrating agent*. Its oxidizing properties are well illustrated by its action on metals, forming sulfur dioxide. The concentrated acid has such a powerful affinity for water (dehydrating agent) that it will extract the hydrogen and oxygen from certain compounds in the proportion of 2 to 1 (as in water). Wood, paper and sugar, for example, are readily charred because of the removal of these two elements, leaving a residue of carbon.

CARBON AND ITS OXIDES

Carbon is an element of tremendous import. Since it possesses the unique chemical property of combining with itself, there is no limit to the possible number and complexity of its compounds. For the moment, however, we are just interested in the element and its oxides. The study of its complex organic compounds will be pursued in the chapter on organic chemistry.

Carbon is extensively distributed in nature. *In the free state carbon occurs as coal, graphite and diamond;* in the combined state it is found in an astronomic number and variety of compounds of plants and animal origin, i.e., organic compounds. Of the free element, there are three definite allotropic forms recognized: *diamond, graphite and amorphous carbon.* Diamond and graphite are *crystalline* varieties. Amorphous carbon is the shapeless form which includes such substances as *coke, charcoal, boneblack and lampblack.* The importance and uses of these various forms of the free element are common knowledge.

Carbon dioxide gas is a vital constituent of the atmosphere, where it is present to the extent of about 0.04 per cent by volume. Important minerals such as marble, limestone and calcite contain carbon dioxide in the form of calcium carbonate. In addition to its ubiquity in nature, carbon dioxide is of interest to us because of its many and varied uses.

PREPARATION

Carbon dioxide is easily prepared in the laboratory by the action of an acid on a carbonate. Usually it is prepared by reacting calcium carbonate with hydrochloric acid. The gas is immediately evolved, and may be collected by displacement of water or by displacement of air. The equation for the reaction is:

$$CaCO_3 + 2HC1 = CaCl_2 + H_2O + CO_2$$

It is to be noted that carbonic acid is first formed. This is an unstable compound, however, and instantly breaks up to form water and carbon dioxide.

Commercially, carbon dioxide is prepared by the burning of carbon (usually in the form of coke) or the heating of limestone. The reactions are as follows:

$$C + O_2 = CO_2$$

PHYSICAL PROPERTIES

Carbon dioxide is a colourless gas, with a slight taste and odor. It is about one and one-half times as heavy as air. Carbon dioxide is moderately soluble in water. Soda water and similar carbonated beverages contain relatively large quantities of the gas held in solution under pressure. The *effervescence* (gas escaping from a liquid) observed when a bottle of pop is opened is due to the release of pressure. Solid carbon dioxide ("snow") is formed when the liquid variety is allowed to escape; the rapid evaporation causes such a sharp drop in temperature that some of the liquid freezes. The useful refrigerant, "dry ice," is made by pressing this snow into a compact form.

CHEMICAL PROPERTIES

Chemically, carbon dioxide is a stable and father inactive gas. It does not burn or support combustion. Carbon dioxide reacts with water to form the weak and unstable carbonic acid and is, therefore, the acid anhydride of that acid:

$$H_2O + CO_2 = H_2C\,O_3$$

PHOTOSYNTHESIS

Carbon dioxide is a participant in probably the most important chemical reaction on earth is *photosynthesis.* This is the term applied to the reaction in which *carbon dioxide* and *water* form simple sugar (glucose) in green plants. Volumes have been written concerning the mechanism involved to accomplish this chemical feat. Although a great deal is known about the intermediate chemical steps, the complete picture is still somewhat nebulous. The biologist, however, is content with the following established facts: *Sugar (glucose) is manufactured in the leaves of plants with the aid of chlorophyll, the complex green pigment of all green plants and sunlight.* Chlorophyll acts as a *catalyst,* and sunlight furnishes the necessary *energy* for the reaction. The following equation summarizes the process:

$$6CO_2 + 6H_2O = C_6H_{12}O_6 + 6O_2$$

CARBON CYCLE

The process of respiration and the process of photosynthesis are mutually sustaining chemical reactions; together they make life on earth possible. The relation between these two processes is commonly referred to as the *carbon cycle.* In this cycle, carbon dioxide is constantly being added to the atmosphere by the processes of respiration, burning and decay. At the same time it is being withdrawn by plants for photosynthesis. Plants utilize carbon dioxide to manufacture food. Oxygen, which is a by-product of this reaction, is given off to the atmosphere, thus allowing animals to carry on respiration. Carbon which is lost during respiration is returned to the animal via his food directly (in plants) or indirectly (other animals). Thus, the carbon cycle involves respiration, photosynthesis and food.

PHYSIOLOGIC IMPORTANCE

Carbohydrates, sugars and starches, and fats are burned within the cells of the body to furnish energy. Since these substances contain carbon, carbon dioxide is given off. Therefore, on a basis of chemistry, respiration may be defined as *the taking in of oxygen and the giving off of carbon dioxide.*

ARSENIC

Arsenic is an *amphoteric* element; i.e., it possesses both metallic and nonmetallic properties. It is a gray solid which resembles a true metal in appearance. Chemically, however, it usually acts as a nonmetal. Arsenic and its compounds are poisonous substances. For this reason they have been employed for years as insecticides and rodenticides. Copper arsenite *(Pans green),* lead arsenate and arsenic trioxide *(white arsenic)* are popular and familiar compounds of arsenic.

In medicine, arsenic has always played an important role. Of special significance is the organic arsenic compound *arsphenamine,* used in the treatment of syphilis. The introduction of arsphenamine commonly referred to as "606" by Ehrlich in 1909 marked a new era in man's conquest of disease. For the first time the physician was given a true *specific* i.e., an agent possessing the ability of destroying an invading pathogenic microorganism. The use of specific chemical agents in disease aptly labeled *chemotherapy* is without question the most outstanding development in medicine. The number of chemotherapeutic agents is ever on the increase. There can be little doubt that the sulfa drugs and the antibiotics (e.g., penicillin, Aureomycin, Terramycin, etc.) are only a preview of things to come.

PHOSPHORUS

Phosphorus is a solid which exists in two *allotropic* forms: *white phosphorus* and *red phosphorus.* These two forms display widely differing properties. White phosphorus ignites at room temperature (it must be kept under water), is soluble in ether and carbon disulfide, and is extremely poisonous. Red phosphorus will not burn until heated, is not soluble in ether and carbon disulfide, and is not poisonous. Red phosphorus is prepared by heating white phosphorus in the absence of oxygen. Both red and white phosphorus in the presence of oxygen are oxidized to *phosphorus pentoxide* the acid anhydride of phosphoric acid:

$$4P + 5O_2 = 2P_2O_5$$

$$P_2O_5 + 3H_2O = 2H_3PO_4$$

Phosphorus is too active to be found in the free state. It occurs chiefly in the form of *phosphates,* i.e., salts of phosphoric acid. Phosphates are found in the soil and in certain phosphate rock deposits, such as in South Carolina and Florida. Phosphorus is an important constituent of the body. *About 80 per cent of the solid matter of bones and teeth is composed of calcium and magnesium phosphates..* The blood, brain, nerve and muscle tissues contain organic and inorganic phosphates.

The principal use of elemental phosphorus is in the making of matches. Originally white phosphorus was used, but due to its high toxicity it has been replaced by red phosphorus. The sodium salts of phosphoric acid are used in medicine. Since phosphoric acid has three replaceable hydrogens, there are three "sodium phosphates": *mono sodium phosphate* (NatLPO,); *disodium phosphate* (Na,HPO_4); and *trisodium phosphate* (Na_3PO_4).

SILICON

Next to oxygen, silicon is the most abundant (28 per cent) element of the earth's crust. Due to the activity of silicon, it is never found free, but in the combined condition viz., as *silica* (SiO_3), *silicates* and *combined silicates. Quartz, zircon, asbestos, talc, mica, clay* and *feldspar* are silicate minerals of extreme importance.

Free silicon is employed almost exclusively in the steel industry. The addition of silicon to low-carbon steel increases its elastic limit and markedly increases its resistance to wear. Of the compounds of silicon, *silica* sand) and *clay* are of chief concern. These two natural compounds are intimately related to the history of man. From sand man makes *glass,* and from clay, *cement.* It is indeed difficult to visualize our civilization without these two materials.

Although silicon is omnipresent and important, it has received little publicity in our chemistry texts. Recent developments (since World War II), however, have drastically changed this practice. A class of compounds called *silicones* offers promise of becoming one of our most valuable classes of compounds. Silicones are complex compounds whose molecules are constituted by silicon and oxygen atoms arranged alternately in chains. By varying the length of the chains, the chemist has been able to synthesize an array of diversified and useful products having the physical properties of oils, rubbers, resins, plastics, and so on. Of special significance is the fact that silicones are probably the most resistant compounds yet synthesized in the laboratory.

Chapter 5

The Metals

Of the 92 natural elements which constitute the earth, more than 70 are metals. Of this number, the great majority are so rare and of such limited commercial value that they are seldom mentioned, even among chemists. Only about 15 metals are commonly encountered in everyday life. Indeed, such an unbelievably complex substance as protoplasm contains no more than 10 metallic elements. In this chapter we shall study the important metals in some detail. The table below summarizes the principal facts and differences relating to metals and nonmetals.

Metals	Nonmetals
Good conductors of heat and electricity	Poor conductors
Unite with nonmetals to form salts	Unite with metals to form salts
Form bases (basic anhydrides)	Form acids (acid anhydrides)
Combine with other elements by electro-Valency	Combine by electrovalency or covalency
form positive ions in solution	form negative ions in solution
During electrolysis are liberated at the cathode	Are liberated at the anode

RELATIVE ACTIVITY OF METALS

When a strip of zinc is placed into a solution of copper sulfate, the strip becomes coated with copper. On the other hand, if a strip of zinc is placed into a solution of some aluminum salt, nothing happens. In the first case, zinc *being more chemically active than copper* replaces the copper from its compound; in the second case, there is no action because *zinc is less active chemically than aluminum.* When a metal replaces another metal from its salt it loses electrons (i.e., it is *oxidized),* while the metal which leaves the solution gains electrons (i.e., it is *reduced).* In other words, the chemical activity of a metal is determined by the ease with which it is oxidized. When the metals are arranged according to their activity, the series is known as the *replacement series* or *electrochemical series.* The electrochemical series is as follows:

1. Potassium (highest activity)
2. Sodium
3. Calcium
4. Magnesium
5. Aluminum
6. Zinc
7. Iron
8. Nickel
9. Tin
10. Lead
11. HYDROGEN
12. Copper
13. Mercury
14. Silver
15. Platinum
16. Gold (lowest activity)

Each metal in this list will replace from its salt solution any metal below it, but will, in turn, be replaced by any metal above it. Hydrogen is included because it acts chemically like a metal. All metals above hydrogen will liberate it from dilute adds, while the metals below hydrogen do not replace it from acids.

HYDROGEN

Hydrogen enjoys the unique distinction of being the simplest and lightest element. Hydrogen is also unique in that it possesses the *chemical properties of a metal,* and *the physical properties of a nonmetai.* In the free state, hydrogen occurs in natural and volcanic gases; traces of the gas are found in the lower strata of the earth's atmosphere. Hydrogen occurs abundantly in the sun's atmosphere. In the combined state hydrogen is ubiquitous; it is present in water, and all animal and vegetable substances.

PREPARATION

Pure hydrogen is prepared commercially by the electrolysis of water: Hydrogen is liberated at the cathode, and oxygen at the anode.In the laboratory, the gas is best prepared by reacting such metals as zinc and iron with certain dilute acids usually sulfuric acid and hydrochloric acid. This reaction is an example of *simple replacement,* where the metal replaces the hydrogen of the acid and releases it as a gas.

PHYSICAL PROPERTIES

Hydrogen is a colourless, odorless, tasteless gas, only slightly soluble in water. Its outstanding property, however, is that it is the lightest gas known. Certain metals, namely, palladium and platinum, have the power to absorb hydrogen in large volumes. This rather strange property is called *occlusion.*

CHEMICAL PROPERTIES

Hydrogen gas burns with a very hot, pale blue flame, producing water. When mixed with oxygen or air it produces a violent explosion on the application of a flame or spark. The equation for the burning of hydrogen is as follows:

When substances which contain hydrogen viz., *fuels* and *foods* are burned, water is always formed. The water vapour in exhaled air is due to the oxidation ("burning") of fats and glucose within the cells of the body. At high temperatures hydrogen combines directly with other elements.

USES OF HYDROGEN

The use of hydrogen in "lighter-than-air" craft has been almost completely superseded by helium. The tragic explosion of the *Hindenberg,* the world's largest dirigible, in May, 1937, served as a final indictment of hydrogen for this purpose. (Helium has about 90 per cent of the lifting power of hydrogen, although it is twice as heavy. Helium has the decided advantage of being noncombustible.) Huge quantities of hydrogen are used in the Haber Process to manufacture ammonia gas. As described before, ammonia is formed by the direct combination of hydrogen and nitrogen, thus: are converted into solids by being made to react with hydrogen. In the laboratory (and to some extent commercially) hydrogen is employed as a *reducing agent,* i.e., *to remove oxygen from certain compounds.* When, for example, hydrogen gas is passed over hot copper oxide, water is formed.

SODIUM, POTASSIUM, AND THEIR COMPOUNDS

The so-called *alkali metals lithium, sodium, potassium, rubidium, cesium and francium* are the most active metallic elements. The activity and similarity of these elements arise from the fact that each of their atoms has a single electron in the outer ring. Reference to the Periodic Table will show that the alkali metals constitute a *family of elements* Family A, Group I. Of the alkali metals, sodium and potassium are by far the most important elements. The others (with the possible exception of lithium) are rare and of slight importance. We shall, then, concentrate on these two elements and their compounds.

Because of their extreme activity, sodium and potassium are never found free in nature. In the combined condition they are widely distributed. The more important compounds of sodium include: *sodium chloride* (common table salt), *sodium nitrate* (Chile saltpeter) and *sodium tetraborate* (borax). The principal natural compounds of potassium include: *potassium chloride* (potash) and *potassium aluminum silicate* (feldspar).

PREPARATION

The preparation of sodium and potassium is essentially the same process today as was used by the great English chemist, Davy. Both metals are prepared by the electrolysis of their fused (melted) hydroxides; sodium and potassium are liberated at the cathode.

PHYSICAL PROPERTIES

These two metals are very similar in appearance. Moreover, their texture is a surprise to one who sees them for the first time; *both are soft enough to be cut with a knife.* The freshly cut surface has a bright metallic luster, but soon darkens on exposure to air. Both are excellent conductors of heat and electricity.

CHEMICAL PROPERTIES

Sodium and potassium react violently with other elements. Their reaction with water or water vapour is so explosive that it must be stored in kerosene. An interesting experiment in the laboratory is to place a *very tiny* piece of potassium or sodium on the surface of water contained in a wide-mouthed receptacle: A reaction ensues instantly, and so much heat is evolved that the metal melts into a globule which dances wildly across the surface of the water. If the piece is large enough, the heat is sufficient to ignite the hydrogen, which is one of the products formed in the reaction. A large piece will produce an explosion. The activity of potassium and sodium with water is a characteristic reaction and an important reaction. In addition to hydrogen gas, the *hydroxide* of the metal is formed:

$$2K + 2H_20 = 2KOH + H_2$$

USES

The *free* alkali metals are not important commercially; in the laboratory they are used only to a limited extent. Their compounds, however, are of extreme importance. The salts of sodium and potassium are by far the most common chemical substances employed either in the laboratory or commerce. Several of these salts deserve special mention.

FLAME TEST

When sodium or a sodium compound is directly heated in the Bunsen burner flame, an intense yellow colour is imparted to the entire flame;* potassium and potassium compounds produce a violet colour. Certain other metals and their compounds also impart characteristic colours to the flame. The colours produced are as follows:

Sodium	yellow
Potassium	violet
Lithium	crimson
Calcium	orange-red
Barium	green
Strontium	bright red

The test is carried out by dipping a clean platinum wire into the substance under investigation (either solid or water solution), and then holding it in the outer edge of the colourless part of the Bunsen flame. Although the test has certain inherent inaccuracies, it is simple to perform, and gives good results when the compound being tested is in the pure form.

SODIUM CHLORIDE (NACL)

Sodium chloride, common table salt, is one of the most valuable compounds known to man. It is found in ocean water, salt beds in the earth and within the cells of living organisms. Students of physiology have a particular interest in this compound.

EXTRACTION

Sodium chloride (usually referred to as "salt") is obtained from salt lakes and ocean water by the simple process of evaporation. The water is piped into shallow basins, and the heat of the sun is allowed to evaporate the water. Salt from salt beds is obtained by mining (similar to coal mining), or by driving pipes into the beds, and allowing water to dissolve the salt. The salt solution formed is then evaporated in the manner described above.

PROPERTIES

Sodium chloride is a white solid made up of cubical crystals. When heated the crystals crackle and explode. This action is called *decrepitation*. Decrepitation is caused by the vaporization of the water which is *mechanically* trapped within the crystal.

In other words, this water is not water of crystallization. Since table salt takes on moisture from the air (especially during humid weather), it would appear that sodium chloride is a deliquescent substance. Pure salt, however, is not deliquescent. This property of table salt results from the presence of deliquescent impurities, namely, magnesium chloride. As for the taste of sodium chloride, we say that it is simply "salty." Other *chemical salts* have a similar taste.

USES

Sodium chloride has many uses other than being our favourite "mineral food." It is used in great quantities to preserve meat and fish. In the chemical industry sodium chloride is the starting chemical for practically all of the important compounds of sodium and chlorine.

SODIUM BICARBONATE

Sodium bicarbonate *(baking soda)* is a common compound, both in the laboratory and home. Of special interest to us is its method of manufacture the *Solvay Process*. The Solvay Process is an example of chemical genius and ingenuity at the industrial level. In practice, the teachings of chemistry mean very little unless pure chemical compounds can be produced in sufficient quantities to supply all needs and *at low cost*. The Solvay Process is herewith described to illustrate the remarkable efficiency which prevails in the chemical industries. In the following steps, note the cheap and abundant raw materials, the use of the by-products to prepare the main product, and the relatively simple and direct chemical reactions.

PROPERTIES AND USES

Sodium bicarbonate is a white solid which dissolves in water; its solutions are mildly alkaline. Sodium bicarbonate is a good source of carbon dioxide, and is employed to generate this gas in the soda-acid fire extinguisher (described earlier) and in baking powders.

In medicine, it is still the most popular agent to neutralize excess acid in the stomach *(hyperacidity)*. The belching following a dose of sodium bicarbonate results from the reaction between the bicarbonate and the hydrochloric acid in the stomach:

$$NaHCO_3 + HC1 \rightarrow NaCl + H_2O + CO_2$$

Sodium carbonate *(washing soda)* is prepared by heating sodium bicarbonate:

$$2NaHCO_3 = Na_2CO_3 + H_2O + CO_2T$$

The above equation applies to all bicarbonates: *They are decomposed into a carbonate, water and carbon dioxide when subjected to heat.* Properties and Uses. When a water solution of sodium carbonate is slowly evaporated, large transparent crystals are formed, having the formula $Na_2CO_3{'}10H_2O$.

When exposed to air this compound (the *hydrous* form) slowly loses its water of crystallization *(efflorescence)*, and crumbles into a white powder (Na_2CO_3) the *anhydrous form.* Solutions of sodium carbonate are basic in reaction, *owing to hydrolysis. For this reason sodium carbonate and sodium bicarbonate should not be confused;* the greater alkalinity of the former compound would cause severe irritation if taken internally. Like all carbonates, it reacts with acids to produce carbon dioxide:

$$Na_2CO_3 + 2HC1 \rightarrow 2NaCl + H_2O + CO_2$$

Sodium carbonate enjoys many commercial uses. It is employed in the manufacture of soap, soap powders, water softeners and glass. In the laboratory it is used whenever a typical soluble carbonate is needed.

SODIUM HYDROXIDE (NAOH)

Sodium hydroxide is made commercially by the electrolysis of a sodium chloride solution *(brine)*. This process (called the *Nelson* or *Vorce Process)* is interesting in that there are two products formed other than sodium hydroxide. It is easy to understand this process if we apply what we already know. The chloride ion (Cl^-), being negative, will be attracted to the anode, and there become *chlorine gas;* the sodium ion (Na^+) will be attracted to the cathode, and become sodium metal.

However, it was pointed out under our discussion of sodium that the metal is extremely active, especially with water. As soon as the sodium ions become sodium atoms, therefore, the atoms react with the water of the solution to form *sodium hydroxide and hydrogen.* The Nelson Process thus enables us to manufacture chlorine, hydrogen and sodium hydroxide simultaneously. In the laboratory the following simple experiment is most instructive: A drop of phenolphthalein indicator solution is added to a beaker of salt solution. When electrodes are placed into this solution and a current applied, bubbles of gas are formed around each electrode. The gas coming off the anode has the odor of chlorine; the gas off the cathode has no odor, but is inflammable hydrogen. Moreover, the area around the cathode turns red (phenolphthalein turns red in basic solution) showing that sodium hydroxide has been formed.

PROPERTIES AND USES

Sodium hydroxide is a white deliquescent solid, freely soluble in water. It is a typical strong base due to the fact that it is highly ionized in solution, furnishing an abundance of hy-droxyl ions. Sodium hydroxide neutralizes all acids, forming sodium salts. It neutralizes hydrochloric acid, for example, forming sodium chloride and water:

$$NaOH + HC1 = NaCl + H_2O$$

Sodium hydroxide is highly caustic to living tissues, dissolving both protein and fat substances. For this reason it is commonly known as *caustic soda.*

In the laboratory sodium hydroxide has many uses. It is, without question, the most frequently used base. Commercially, sodium hydroxide is employed in many ways. Huge quantities are consumed in the manufacture of soap from fats. Most everyone is familiar with the word *lye.* Lye is a commercial form of sodium hydroxide used to clean toilets and sink drains. Lye cleans by attacking and dissolving organic matter especially fats, oils and grease.

SODIUM NITRATE

Huge deposits of sodium nitrate, known as *Chile saltpeter,* are found in Chile and Peru. These deposits were formed by a "natural coincidence," since *organic matter, sodium salts and dry air* must be present. For years the world was dependent upon these deposits for a source of cheap nitrate.

PROPERTIES AND USES

Sodium nitrate is a white crystalline solid, very soluble in water. Great quantities of sodium nitrate are used in the manufacture of fertilizer and nitric acid. The nitric acid, in turn, is employed to manufacture a greater number of organic and inorganic nitrate compounds. In our discussion of artificial nitrogen fixation as applied to ammonia and nitric acid, we learned that man no longer depends upon natural nitrogen compounds, namely, the Chile deposits. At present more than half the world's needs are supplied by artificial fixation, viz., the *Haber Process and the Synthetic Ammonia Process.*

POTASSIUM COMPOUNDS

Compounds of potassium, although frequently encountered in the laboratory, pharmacy and commerce, are decidedly less common than those of sodium. Probably the chief reason for this is that potassium is more expensive than sodium. Chemically and physically, sodium and potassium salts are almost identical in their behaviour. Physiologically, however, they are not similar; protoplasm readily sustains an insult when a K+ ion is substituted for a Na^+ ion, and vice versa. The body, for example, would soon show signs of metabolic dysfunction if potassium chloride were to be substituted for sodium chloride. Some of the common and important compounds of potassium are herewith briefly described.

POTASSIUM CHLORATE ($KC1O_3$)

Potassium chlorate is a white crystalline solid, possessing strong powers of oxidation. As with all powerful oxidizing agents, great care should be exercised in its use. Potassium chlorate is a valuable oxidizing agent both in the laboratory (it is used, you will remember, to prepare oxygen) and in industry where it is employed to manufacture matches, fireworks and explosives.

Potassium Chloride (KC1). Potassium chloride occurs in the minerals *carnallite* ($KCl\text{-}MgCl_2\text{-}6H_2O$) and *sylvite* (KC1). These two minerals are obtained from mines at Strassfurt, Germany. Potassium chloride is probably, by virtue of its relative abundance, the most important compound of potassium. It is used as a fertilizer, and in the preparation of other potassium compounds. Potassium is an essential element to plants.

Potassium Hydroxide (KOH). Potassium hydroxide, commonly known as *caustic potash,* is physically indistinguishable from sodium hydroxide. Moreover, it is prepared by the same process *the electrolysis of potassium chloride.* Potassium hydroxide is probably the second most common base in the laboratory; industrially, however, it has a rather limited usefulness. Of interest is the fact that *soft soaps* are made from the reaction between fats and potassium hydroxide; *hard soaps* are made from solid fats and sodium hydroxide.

POTASSIUM NITRATE (KNO_3)

Potassium nitrate, commonly known as *saltpeter (Chile saltpeter is sodium nitrate),* is of some importance. It is a white crystalline solid, freely soluble in water. Chemically, it is a rather potent oxidizing agent, and is sometimes employed as an oxidizing agent in the laboratory. Commercially, it is used as a fertilizer, meat preservative and in the manufacture of gunpowder. Gunpowder, which has probably directed the destiny of man more than any other substance, is a simple mixture of *potassium nitrate, sulfur and charcoal.* In medicine, potassium nitrate has been used as a diuretic (to increase production of urine) and as a hypotensive agent, i.e., to reduce blood pressure.

CALCIUM AND ITS COMPOUNDS

Calcium is the principal member of the *alkali-earth metals.* The alkali earth metals are a family of elements of Family A, Group II of the Periodic Table. The family includes the following elements: *calcium, beryllium, magnesium, strontium, barium and radium.* The physical and chemical properties of these elements and their compounds are very similar.

Next to aluminum and iron, calcium is the most abundant metal of the earth's crust. Since it is chemically reactive, it occurs in a number of compounds. The most abundant and important calcium salts in nature are *calcium carbonate, calcium suljate, calcium fluoride and calcium phosphate.* In the body, calcium is by far the most important metallic element; the bulk of *bones and teeth* is constituted by calcium in the form of *calcium phosphate.*

PROPERTIES AND PREPARATION

Calcium is a light, silver-white metal which tarnishes rapidly in moist air. It reacts with oxygen to form calcium oxide, and with water to form calcium hydroxide and hydrogen. Calcium metal is prepared by the electrolysis of calcium chloride.

Uses. Calcium in the free state has few uses. Its compounds, however, are of vast importance in the laboratory, medicine and industry. The more important calcium compounds are herewith described.

CALCIUM CARBONATE ($CACO_3$)

Calcium carbonate is the most abundant calcium compound in nature. Of considerable interest is the fact that calcium carbonate occurs as *marble, limestone, chalk and calcite.* The mineral calcite is the only crystalline variety of calcium carbonate all the other forms are amorphous solids.

Properties

Calcium carbonate is a white amorphous solid, insoluble in water. It is decomposed by heat into calcium oxide and carbon dioxide;

$$CaCO_3 = CaO + CO_2$$

Like other carbonates, it reacts with acids to form salts and carbon dioxide gas:

$$CaCO_3 + 2HC1 = CaCl_2 + H_2() + CO_2$$

LIMESTONE CAVES

In regions where there are large deposits of limestone, the slow solvent action of soil water has formed tremendous caves of weird and awesome beauty. The solvent action of soil water on limestone is due to its acidity; the acidity, in turn, is due to the presence of carbon dioxide which rain water dissolves on its journey through the air.

Calcium bicarbonate ($Ca(HCO_3)_2$) is a *soluble compound,* and is removed by the ground's water. A singular feature of limestone caves is the icicle-like projections from their roofs and floors. These are easily explained by the fact that calcium bicarbonate readily decomposes into calcium carbonate, carbon dioxide and water. Therefore, when water containing calcium bicarbonate drips from the roof of a cave, some of the water evaporates, leaving $Ca(HCO_3)_2$, which subsequently decomposes as follows:

$$Ca(HCO_3)_2 = CaCO_3 + H_2O + CO_2$$

These downward projections are called *stalactites.* The upward projections, or *stalagmites,* are formed in the same way from the water which has dripped to the floor. In time, the stalagmites meet the stalactites to form columns. Eventually (thousands of years) this process will fill the cave up again. Famous limestone caves in the United States include:

Mammoth Cave (Kentucky), the Luray Caverns (Virginia), Carlsbad Caverns (New Mexico) and Howe Caverns (New York State).

USES

Calcium carbonate in its natural forms (marble and limestone) has served the building needs of man for thousands of years. Limestone is consumed in huge quantities to make glass, lime and cement. In the blast furnace, limestone is employed in the extraction of iron from its ore. Pure calcium carbonate (commonly referred to as *precipitated chalk)* is used to make tooth powder, toothpaste and polishes. Calcium carbonate has proved to be an excellent medicinal agent in hyperacidity. In the laboratory it is often used to prepare carbon dioxide.

Other chemical substances which react in a similar manner are sold on the market. On a large scale, *permutit* is probably the most efficient water softener. Permutit is a complex chemical compound which contains replaceable sodium atoms. When water is allowed to pass through a tank of this substance, the calcium and magnesium ions in the water are *exchanged* for sodium ions; sodium salts in water do not react with soap. Another approach to the hard water problem was the advent of the so-called *detergents (Tide, Dreft, Vel, Fab, Cheer, etc.).* Detergent soaps are really not soaps; their molecular configuration is decidedly different. Being different, they do not mimic the action of soap toward hard water. Indeed, they clean as well in hard as soft water. Their general superiority to conventional soap is fully established.

IRON AND ITS COMPOUNDS

Iron is the second most abundant metal of the earth's crust (aluminum is first). It occurs chiefly in the form of oxides. *Hematite,* for example, an important iron ore. In the biologic world, iron in the form of *hemoglobin* is found in the blood of nearly all animals. Iron is of vast importance to man. In the form of steel, it has revolutionized our civilization. In addition, its chemical compounds have hundreds of diversified uses in the laboratory, commerce and medicine.

EXTRACTION

Iron is obtained from its oxide ores by the chemical process of *reduction,* i.e., the removal of oxygen. The reduction or extraction process is carried out in the *blast jurnace* at tremendously high temperatures. The blast furnace is one of the most impressive sights on the industrial scene. The typical furnace is about 100 feet high and 25 feet in diameter. It is built of wrought iron and is lined with fire brick. The furnace is charged with the following materials:

(1) *iron ore,* usually hematite (2) *coke,* a porous form of carbon, to supply heat and act as a reducing agent; (3) *hot air,* to burn the coke; and (4) *flux,* usually limestone ($CaCO_3$), to combine with the impurities in the ore. When the furnace is started, a series of chemical reactions take place between these materials to produce molten iron. The first reaction is the combustion of the coke near the bottom of the furnace. The heat resulting from this reaction renders the upper layers of coke red-hot. As carbon dioxide formed in the combustion of the coke passes up through the heated mass, it is reduced to carbon monoxide. Being a powerful reducing agent at high temperatures, the carbon monoxide removes oxygen from the iron ore, forming iron and carbon dioxide.

At the same time some ore is directly reduced by the hot coke in the lower part of the furnace. The flux calcium carbonate is decomposed by the intense heat of the furnace, forming calcium oxide and carbon dioxide. The calcium oxide unites with the impurities of the ore chiefly *sand and quartz* (SiO_2) forming a glasslike *slag,* which has the formula CaSiO, (calcium silicate). At the bottom of the furnace, the molten slag floats on the surface of the molten iron. It is ther removed through the "slag notch." Molten iron is run out through the "iron notch."

The waste gases resulting from the reactions in the furnace pass out through the pipe located at the top. Since they contain about 25 per cent carbon monoxide, they are used to heat the air which is blown into the furnace.

TYPES OF IRON

Cast Iron

Cast iron, or *pig iron,* is the iron obtained from the blast furnace. It is the least pure of all forms of iron, containing about 5 per cent carbon and 2 per cent of sulfur, silicon, phosphorus and manganese. Cast iron is a hard, brittle solid, with a crystalline structure. It melts in the vicinity of 1200° C. Seventy-five per cent of all the cast iron from the blast furnace goes into the making of wrought iron and steel. The remainder is *cast* into various articles such as radiators, stove parts and water pipes, i.e., articles which are not intended to sustain stress.

Wrought Iron

Wrought iron is the *purest form of commercial iron,* containing less than 0.5 per cent foreign matter. It is made by oxidizing the impurities in cast iron. The process is carried out in a *reverberatory furnace.* Wrought iron in contrast to cast iron is tough, flexible, malleable and ductile. It cannot, however, be tempered. Wrought iron is the principal type of iron used in making pipes, boiler tubes, chains,- anchors, bolts and electromagnet cores.

STEEL

Steel is, without question, the most important form of iron. It can be tempered, forged into any desired shape, cast into molds, rolled into bars and sheets, and drawn into wire. By virtue of the variety of ways it may be worked, steel is our most useful metallic material of construction.

Chemically, steel is a form of iron intermediate in purity between cast iron and wrought iron. The various types of steel contain from 0.2 per cent to 2 per cent of carbon. It is a most interesting fact that the type of steel depends on the *concentration of the various impurities especially carbon.* In contrast to cast and wrought iron, therefore, the making of steel demands strict analytic control. Essentially, the making of steel involves the removal of impurities from cast iron, and the subsequent addition of *measured amounts of carbon* and / or other metals.

The *Bessemer Converter* an early, but still important process illustrates the method. About 15 tons of molten cast iron direct from the blast furnace are placed in an egg-shaped converter. A strong blast of heated air is then blown through the liquid to bring about the oxidation of carbon and other impurities. After about fifteen minutes, the oxidation is complete, as indicated by the size and colour of the flame issuing from the mouth of the converter. At that time, weighed quantities of carbon and manganese are added to provide the correct amount of foreign matter.

Bessemer steel is only one variety of steel. At the present time most of our steel is made by the *Open-Hearth Process.* Open-hearth steel is of a much higher grade than Bessemer. Other important processes for making steel include the *Crucible Process* and the *Electric Furnace Process.* Special steels are produced when a certain property is desired, such as unusual hardness, toughness and chemical resistance. Special steels i.e., steel alloys arc produced by adding small amounts of other metals to ordinary steel. Chrome steel, for example, is the steel of choice where resistance to corrosion is desired. Chrome steel contains 3 per cent chromium.

COMPOUNDS OF IRON

When the atoms of iron combine with other elements, they are capable of losing 2 or 3 electrons, forming the ions $Fe^{4'}$ or Fe^{4++}. In other words, an atom of iron has 3 electrons in its outer orbit. Whether it loses 2 or 3 electrons depends upon the nature of the reaction, and the element with which it is combining. Therefore, we possess two types of iron compounds: those in which iron has a *valency of* 2, and those in which iron has a *valency of 3.* Those in which iron has a valency of 2 are called *ferrous* compounds; those in which iron has a valency of 3 are called *ferric* compounds. In general, ferrous salts are green, while ferric compounds are yellow, brown or red. It is most important to understand that ferrous compounds are not as stable as ferric compounds. This is to be expected, since a ferrous ion is capable of losing an additional electron. Ferrous compounds, then, especially when exposed to air for long periods of time, gradually change to ferric compounds. In the laboratory, therefore, ferrous compounds should always be freshly prepared in order to prevent "ferric contamination."

How do we know whether a given iron compound is in the ferrous or ferric condition? A very simple, specific, and highly sensitive test may be employed to differentiate the two types of compounds. When a dilute solution of *potassium ferricyanide* is added to a ferrous salt solution, a beautiful blue precipitate, called *Turnbull's blue,* forms immediately. Ferric salts, ferric sulfate, for example, react immediately with *potassium ferrocyanide* ($K_4Fe(CN)_6$) to form an intense dark blue precipitate called *Prussian blue.*

There are quite a few important compounds of iron both ferrous and ferric in the laboratory, industry and pharmacy. The following compounds are common in any chemical laboratory, and serve to illustrate the chemistry of iron.

Iron Oxides. From what was stated above, we can readily understand that iron forms two oxides with oxygen: *ferrous oxide* (FeO) and *ferric.* Ferrous oxide is an unstable black powder which readily takes on more oxygen from the air to form ferric oxide. Ferric oxide occurs abundantly in nature as the all important ore, *hematite.* Ferric oxide is a reddish compound. It is the chief compound formed as a result of the rusting of iron. Pure ferric oxide is employed as a paint pigment *(Venetian red),* and as a polishing powder *(jeweler's rouge).*

Iron Hydroxides. When a ferrous salt is treated with ammonium hydroxide, a fluffy white precipitate is formed. On standing a few minutes, the white colour changes to a grayish-green colour, and finally to a reddish-brown colour.

The white precipitate is *ferrous hydroxide,* the reddish-brown compound is *ferric hydroxide.* This experiment illustrates not only the mode of preparation, but the unstable chemical nature of ferrous compounds.

IRON CHLORIDES

When an excess of iron is reacted with hydrochloric acid, *ferrous chloride* is formed. Ferrous chloride, being extremely unstable, changes to *ferric chloride.* Ferric chloride is a reddish-brown compound, quite soluble in water. *Tincture of ferric chloride* (ferric chloride dissolved in alcohol) has been used for years as a *hemostatic agent,* i.e., to arrest bleeding. In the laboratory ferric chloride serves the chemist as a generally useful iron compound.

IRON SULFATES

Ferrous sulfate ($FeSO_4$) is probably the most common compound of iron. The hydrous, or crystalline, form ($FeSO_4'7H_2O$) is known as *copperas* or *green vitriol.* When ferrous sulfate is heated it loses its water of crystallization and, as a result, changes to an impalpable white powder. Provided ferrous sulfate is properly storedÿÿout of contact with air and light, it is relatively stable. Ferrous sulfate finds use in making ink, tanning, and as a dye mordant. It is to be recalled that in the laboratory freshly prepared ferrous sulfate is employed to identify the nitrate radical. In medicine, ferrous sulfate is the *compound of choice* to supply iron to the body. Upon oxidation, ferrous sulfate is converted into ferric sulfate ($Fe_2(SO_4)_3$). Ferric sulfate is used to prepare a class of compounds called *alums.*

Hemoglobin. Without question, hemoglobin is the most singular compound of iron in the biologic world. For the present, let us appreciate that hemoglobin is a highly complex substance whose molecules contain an atom of iron in combination with a protein called *globin.* Hemoglobin is the compound which imparts the characteristic red colour to blood. Another fact to be recalled is that hemoglobin is contained in the erythrocytes or red blood corpuscles; none is found in the plasma. In our future discussions of blood chemistry, we shall have occasion to talk at some length on the chemistry of hemoglobin. For now, let us simply state the all-important function of hemoglobin: *to carry oxygen to the cells.*

As the blood passes through the lungs, the hemoglobin molecules of the red cells combine by virtue of their affinity for oxygen with oxygen to form *oxyhemoglobin.* Since oxyhemoglobin is an unstable compound, oxygen is released in the tissues. The relation of oxygen and hemoglobin may be summarized in a simple way as follows:

Hemoglobin + O_2= oxyhemoglobin (lungs)

Oxyhemoglobin → hemoglobin + oxygen (tissues)

ALUMINUM AND ITS COMPOUNDS

Aluminum is the most abundant metal of the earth's crust. It is present to the extent of about 7 per cent. Since aluminum possesses a high degree of chemical activity, it is not found in the free state. Perhaps this explains why most people are surprised to learn that aluminum is our most common metal. In the combined state, it is present in the form of complex silicates such as *clay, feldspar* and *mica.* Aluminum is also found in *bauxite* and *cryolite.*

PREPARATION

The tremendous value of aluminum stems from the ingenious work of Charles M. Hall. Prior to the so-called *flail Process* for extracting aluminum from its ore, aluminum metal was on an economic par with gold. It was not until 1886 that the earth's reservoir of this valuable metal yielded to the efforts of man. In that year Hall, while still a student at Oberlin College, Ohio, devised an electrolytic method of extraction. The Hall Electrolytic Process, essentially the same as the original, employs *pure bauxite* and *melted cryolite* (which acts as a solvent). The aluminum oxide dissociates (ionizes) in solution to form aluminum ions ($A1^{4++}$) and oxygen ions (O^{-}). When a direct current of electricity is passed through the solution, the *aluminum ions are attracted to the cathode,* and *the oxygen ions to the anode.* Because of the high temperatures of the electrolytic cell, molten aluminum forms at the bottom.

PROPERTIES

Aluminum is silvery-white in colour, and is the lightest of the common metals. It possesses great tensile strength. Aluminum is malleable, ductile, and an excellent conductor of heat and electricity.

Chemically, aluminum is rather active. It readily reacts with oxygen to form aluminum oxide. Aluminum reacts with acids especially hydrochloric acid to form salts and hydrogen.

Aluminum, *Unlike most metals,* is *dissolved by strong bases.* In this reaction, hydrogen and a class of salts called *aluminates* are produced.

USES

The Hall Process, as indicated above, drastically reduced the price of aluminum. This made it possible for industry to develop fully and put to use the wonderful and outstanding properties of this metal. Indeed, aluminum is a *sine qua non* to aviation and the culinary arts. At the present time, automotive and aircraft construction provide the largest single market for aluminum. When used for construction purposes, aluminum alloys e.g., *duralumin* (Al, Cu, Mg), *aluminum bronze* (Al, Cu) and *magnalium* (Mg, Al) arc preferred to the plain metal. Huge quantities of aluminum are employed in the manufacture of kitchen utensils, where its lightness and freedom from tarnish are unsurpassed. Large quantities of aluminum are also used in the making of electric transmission lines and aluminum foil.

Since aluminum has an affinity for oxygen, it is used as a reducing agent. When mixed (in the form of a powder) with certain metallic oxides and ignited by a magnesium fuse, the aluminum removes oxygen from the oxide. Such an enormous amount of heat is evolved that the metal is produced in the molten state. *Thermite* (the name applied to a mixture of aluminum and ferric oxide (Fe_2O_3)) is used for emergency welding. In this procedure, the thermite powder is placed in a heat resistant, funnel-shaped receptacle which is held over the parts to be welded. When ignited by a piece of magnesium ribbon (magnesium burns with the evolution of tremendous heat) the mixture reacts, and molten iron flows out from the bottom. The chemical reaction of thermite is as follows:

$$2Al + Fe_2O_3 = 2Fe + Al_2O_3$$

Chromium, manganese and other metals are extracted from their oxide ores by this so-called *thermite reaction.*

ALUMINUM OXIDE (AL_2O_3)

Aluminum oxide exists in several physical forms in nature. These forms include *bauxite, corundum, emery, rubies* and *sapphires.* All of these forms are of extreme importance. Bauxite is the chief ore of aluminum; corundum and emery are extremely hard, and are used as abrasives.

Rubies and sapphires are precious stones of characteristic beauty. The colour of these gems is due to minute quantities of impurities.

ALUMINUM SILICATES

Complex compounds of aluminum with *silicon, oxygen* and other elements are referred to as *silicates.* Most of the earth's aluminum is present m silicates. *Clay,* an impure aluminum silicate, is found in varying amounts in all soils save pure sand. Huge quantities of clay are employed in the manufacture of *ceramics* (porcelain, china, earthenware) and *cement.*

Pottery is of special concern to the archaeologist; it was used by man long before the dawn of written history. Indeed, the making of pottery was one of man's first talents. While pottery marks in a sense the dawn of man, cement epitomizes (like steel) his present status. It would indeed be difficult to visualize our present society without cement structures. Perhaps even more philosophical is the idea that man has taken clay earth, if you please and from it molded his civilization.

OTHER IMPORTANT METALS

MAGNESIUM

Magnesium is a light, silvery-white metal. It is very active chemically, and burns with a brilliant white light, forming magnesium oxide (MgO). Magnesium readily reacts with acids, forming hydrogen and salts. Magnesium metal is used chiefly in the production of light alloys. *Duralumin,* for example, an extremely tough and light alloy, is composed of aluminum, copper and magnesium. Another important alloy called *magnalium* (an alloy of aluminum and magnesium) is extensively employed in aircraft construction. Magnesium is also used in making fireworks and flashlight powders. Several compounds of magnesium have important uses and applications. The following compounds are of interest to us because of their extensive use in medicine.

MAGNESIUM HYDROXIDE

When a base is added to a magnesium salt, a light, white precipitate of magnesium hydroxide forms immediately. Magnesium hydroxide is used extensively as an *antacid* to neutralize excess hydrochloric acid in the stomach (hyperacidity). The reaction is one of simple neutralization. Larger doses of magnesium hydroxide produce catharsis. In the pharmacy, magnesium hydroxide is sold in tablet form and liquid form. The liquid form of magnesium hydroxide is a white viscous suspension of the compound in water. The preparation is known commonly as *milk of magnesia.* In pharmacy, it is referred to as *magnesia magma.*

MAGNESIUM SULFATE ($MGSO_4$)

Crystalline magnesium sulfate has the formula MgSO, $7H_2O$. The anhydrous form, of course, has the formula $MgSO_4$. Most people know magnesium sulfate by its popular name, *Epsom salt.* Epsom salt is without doubt the most popular saline laxative. It is also a potent anticonvulsant when administered by injection into the body. Larger doses produce anesthesia. Since magnesium sulfate is not absorbed from the intestine, there is no anticonvulsant and anesthetic action followi administration. Orally, then, magnesium sulfate produces cathartic action.

MAGNESIUM OXIDE (MGO)

Magnesium oxide is a white, light and fluffy powder, insoluble in water. It is formed when magnesium ribbon is burned. Magnesium oxide enjoys some use as a laxative and antacid.

MAGNESIUM CARBONATE

Magnesium carbonate is a white, light and fluffy powder, insoluble in water. Like magnesium oxide, it enjoys some use as an antacid.

MAGNESIUM TRISILICATE

Magnesium trisilicate is a compound of rather indefinite molecular constitution. The formula has been assigned to this substance. Magnesium trisilicate is an excellent antacid. Being very insoluble, it affords sustained antacid action. In addition, magnesium trisilicate is bland, and somewhat of a demulcent. Several antacid preparations in the pharmacy contain this compound alone, or in combination with other neutralizing agents.

MERCURY

Mercury possesses the unique distinction of being the only metal that is liquid at room temperature. Mercury is a very heavy silvery-white metal. Chemically, mercury is interesting because it forms two series of compounds: *mercurous,* in which the mercury atom displays a valency of a *plus 1* (Hg^+); and *mercuric,* in which the mercury atom has a valency a *plus* 2 .In this connection and with reference to the valencys of iron, note that when a metal (like mercury and iron) possesses two valencys, the suffix *-ous* is attached to the state of the lower valency *(mercurous, ferrous),* and the suffix -/r to the higher valency *(mercuric, ferric).* Like iron, the "double" valency of mercury in its compounds should be thoroughly understood and appreciated. This is dramatically illustrated by the two chlorides of mercury: *mercurous chloride* (HgCl) and *mercuric chloride* . *Mercurous* chloride may be taken internally; mercuric chloride is *extremely poisonous.*

Mercury metal is particularly well adapted for use in thermometers, barometers, and many electrical devices. Mercury alloys, called *amalgams,* are also of importance. Perhaps the most important amalgam is *silver amaigam,* employed in dentistry for filling tooth cavities. In ancient times, metallic mercury was used to treat disease. There are many compounds of mercury employed commercially and in medicine. Those compounds of medicinal value arc of special interest to us. The following compounds of mercury "often referred to in the healing arts as *mercurials* are in common use today.

MERCUROUS CHLORIDE

Mercurous chloride is a heavy white powder, insoluble in water. It is prepared by reacting any soluble mercurous salt with hydrochloric acid. Mercurous chloride has enjoyed wide use as a rather potent laxative. In pharmacy and medicine it is commonly known as *calomel.*

MERCURIC CHLORIDE

Mercuric chloride is usually referred to in pharmacy and medicine as *bichloride of mercury* or *corrosive sublimate.* It is a heavy, odorless, crystalline or white, powdery substance, soluble in water. In contrast to *mercurous chloride,* mercuric chloride is a potent poison. Because of its wide use in medicine and industry, mercuric chloride is responsible for many deathsýÿboth accidental and suicidal. Solutions of mercuric chloride are useful for the disinfection of the skin, instruments, excreta and discharges. Solutions of from 1:100 to 1:10,000 are generally employed. A solution of mercuric chloride (1:1,000) in 25 per cent alcohol is a very efficient germicide.

MERCURIC OXIDE (HGO)

Mercuric oxide is a yellow-orange, heavy impalpable powder. It is insoluble in water. This compound of mercury is employed almost exclusively in medical practice as *yellow mercuric oxide ointment,* which contains 1 per cent HgO. This preparation is used mainly for application to the eye as a mild antiseptic in conjunctivitis, blepharitis and styes.

MERCURY AMINO CHLORIDE

Mercury amino chloride is commonly known in pharmacy as *ammoniated mercury.* It is a white, amorphous powder, insoluble in water. Ammoniated mercury is prepared by reacting mercurous chloride with ammonium hydroxide. *Ammoniated mercury ointment* is successfully employed in the treatment of numerous skin lesions.

ORGANIC MERCURIAL COMPOUNDS

At this point it would be premature to enter into a discussion of metallic derivatives of organic molecules. We shall study this interesting subject under organic chemistry. However, it can be stated that certain metallic organic compounds are of prime importance in certain areas of medicine.

This is particularly true of mercury. Several organic mercurials are employed today. *Mer-curochrome, Merthiolate* and *Metaphen* are popular antiseptics containing mercury in organic combination. The antiseptic potency of the inorganic salts of mercury, viz., bichloride of mercury, suggested the synthesis of less toxic organic compounds. They are less toxic because they do not undergo ionization for it is the mercury ion which is toxic.

Several organic mercurials have been synthesized for use as *diuretics,* i.e., agents which increase the formation of urine. The salts of mercury have been known for many centuries to elicit diuretic response. However, as explained above, inorganic salts due to extensive ionization are toxic. The modern era of the mercurial diuretic began with the use of organic compounds. The first organic mercurial, *Mersalyl,* was introduced in 1924. Soon other mercurials followed with even lower toxicitv *Mercurophylline, Mercuhydrin* and *Thiomerin* are three popular such agents. Mercurial diuretics appear to be the most potent and dependable of all diuretics. In edema of cardiac origin, they are the drugs of choice.

COPPER

Due to the fact that copper occurs rather abundantly in the free state, it was the first metal which primitive man learned to use as a material for tools and weapons; *this was the first step forward from the Stone Age.* The passing of time has not diminished the importance of this reddish metal indeed, centuries have *enhanced* the role of copper. Our present economic life would be paralyzed without copper, for this metal is the most logical medium for the transmission of electric currents. More than half the world's supply of this vital metal is produced in the United States. There are rich deposits of free copper in northern Michigan. In the combined state, the principal ore, *chalcopy-rite,* is found in Montana and Utah.

Copper is a heavy metal with a characteristic reddish colour. Next to gold and silver, it is the most malleable and ductile of the common metals. As a conductor of electricity, it is surpassed only by silver. In air copper oxidizes slowly, forming a protective coating which prevents further corrosion. This green protective coating, called *basic copper carbonate* results from the action of the moisture, oxygen and carbon dioxide present in the air. In the electrochemical series, copper is below hydrogen; therefore it *cannot* displace the hydrogen of acids. Dilute acids have little effect on copper; concentrated acids, however, of the oxidizing type, such as sulfuric and nitric, readily attack copper.

COPPER SULFATE

Copper sulfate is the most important salt of copper. The hydrous form, is composed of beautiful deep blue crystals. Upon heating the crystals 5 molecules of water are driven off, leaving a white anhydrous powder. When water is added to this white powder, a blue colour develops; i.e., the hydrous, or blue, variety is re-formed. Copper sulfate or, more correctly, cupric sulfate is employed in electric batteries, as a mordant in dyeing, and in copper plating. In medicine, copper sulfate is used as an astringent and emetic; it is also used as an antidote in phosphorus poisoning. Benedict's solution, used for testing the presence of sugar in urine, contains copper sulfate. This important test will be thoroughly presented when we discuss sugars.

SILVER

Silver is *the most common of the precious metals.* It occurs in nature both in the free and combined state. Silver is a soft, white metal living a high luster. Next to gold, it is the most ductile and malleable metal. Silver possesses the distinction of *surpassing* all metals as a conductor of electricity.

Chemically, silver metal is inert to oxygen. Toward sulfur, i.e., sulfur compounds, however, it reacts to form a black compound, *silver sulfide* (Ag_2S). The tarnishing of silver is due to the formation of deposits of this substance. Since sulfur compounds are present in the air and in certain foods (such as eggs and mustard), silver utensils require frequent polishing a fact every housewife knows. Silver salts are prepared by reacting the metal with the appropriate concentrated acids.

Dilute acids do not attack silver. Silver nitrate, for example, is prepared by reacting silver with concentrated nitric acid. Since silver is *below* hydrogen in the electrochemical series, no hydrogen gas is released.

Because silver occurs in the free state, man at an early date was fascinated by its beauty and utility. Silver is used in coins and jewelry, and as a coating on silver-plated ware. So-called "solid silver" contains a certain percentage of copper to increase its hardness. Silver coins, for example, contain 90 per cent silver and 10 per cent copper. Such silver is said to be *900 flue* (900 parts of silver in 1000). Sterling silver is *925 fine.* Silver compounds are also important. The following compounds are of special importance.

SILVER NITRATE

Silver nitrate *(lunar caustic)* is the most important compound of silver. It not only possesses valuable qualities, but also is used to prepare other silver compounds. Silver nitrate has been used in medicine for years as a caustic (to cauterize wounds), and as an antiseptic. Silver nitrate in a solution of 1:1000 inhibits microorganisms. For the prevention of *gonorrheal conjunctivitis* in the newborn, a drop of 1 per cent solution is instilled into each eye as soon as practicable after delivery.

SILVER BROMIDE (AGBR)

When a silver nitrate solution is added to a solution of a bromide, a copious white precipitate of silver bromide is formed, thus:

$$AgNO_3 + KBr = KNO_3 + AgBr$$

Enormous quantities of silver bromide are used each year in photography. Today the photographic arts and sciences have grown to giant proportions, consuming about *200 tons of silver annually.*

SILVER PROTEIN

Silver protein is a dark brown or almost black shining crystalline *(scales)* or granular substance. It has been used in medicine for many years as an antiseptic and demulcent. Silver protein is prepared by reacting silver nitrate with a protein substance; the silver atoms "add on" to the large protein molecules. A popular form of silver protein on the market is sold as *Argyrol,* a dark aqueous solution of silver protein. Silver protein preparations in contrast to the inorganic salts, viz., silver nitrate are not irritating to the skin or mucous membranes.

LEAD

Lead is the heaviest of the common metals. It is unusually soft and malleable. When lead is freshly cut it has a bright luster, but soon dulls due to the formation of a protective coating. Nitric and acetic acids readily attack lead, but hydrochloric and sulfuric acids have little action.

Lead is used extensively in the manufacture of water pipes, sheeting and certain alloys. Its chief uses, however, are in making lead plates for storage batteries, and *white lead for paints.* White lead is the most important compound of lead. White lead is a heavy white powder which serves as our most popular paint *base.* A "complete paint" consists of a *base,* usually white lead; a *vehicle,* usually linseed oil; a *thinner,* usually turpentine; a *"dryer,"* i.e., a chemical which catalytically hastens drying (e.g., lead oxide); and a *pigment* to give colour.

LEAD POISONING

More has been written about the toxicology of lead than any other substance. Lead poisoning *(plumbism)* is as old as the history of man. In spite of the tremendous growth of the chemical industry, *lead is still the leading industrial poison.* Lead poisoning is almost always chronic, and occurs among workers in a host of industries where lead metal is employed. The signs and symptoms are insidious because of their gradual onset and their similarity to other abnormalities. Modern methods of industrial hygiene have done a great deal in reducing the incidence of this form of poisoning. Treatment for *plumbism* is largely symptomatic.

BISMUTH

Bismuth is a grayish-white metal which occurs both in the native and combined state. Bismuth metal is used principally in the preparation of alloys with low melting points, such as *Wood's* metal. This alloy is used in automatic sprinklers, electric fuses and other safety devices. Several salts of bismuth are employed in medicine. The following compounds are important.

BISMUTH SUBCARBONATE

Bismuth subcarbonate is a white powder, insoluble in water. This compound is used extensively in treating conditions characterized by irritation along the alimentary tract. It forms a coating over an irritated or ulcerated surface, and thus protects against mechanical and chemical injury. Bismuth subcarbonate is also employed, for the same reasons, on the skin. Bismuth subnitrate is sometimes used for the same purposes. The subcarbonate, however, is the preferred compound.

BISMUTH POTASSIUM TARTRATE

At one time bismuth salts were used routinely usually along with the salts of mercury and arsenic as *antisyphilitics.* The advent of penicillin, however, has all but replaced "heavy metal" therapy. In the event a soluble bismuth salt is desired, bismuth potassium tartrate is usually preferred. This salt is administered by intramuscular injection.

BISMUTH SUBSALICYLATE

Bismuth subsalicylate is a white amorphous powder, insoluble in water. As an antisyphilitic, it is sometimes injected intramuscularly as a suspension in oil. Being insoluble, the subsalicylate provides sustained action, in contrast to the above compound, which gives a prompt bismuth effect. Orally, bismuth subsalicylate, like the subcarbonate, acts as a protective and an astringent.

BISMUTH SUBGALLATE

Bismuth subgallate is an amorphous, bright yellow powder, insoluble in water. Its action is similar to that of the subcarbonate. It is used rather commonly as a *dusting powder* to protect the skin and promote healing.

ZINC

Zinc is a bluish-white metal which reacts with the air, forming a protective coating. The metal occurs in the combined state, chiefly as the sulfide (ZnS). Zinc readily reacts with most dilute acids, liberating hydrogen gas and forming zinc salts. Zinc metal is used for roofing, lining tanks, and coating iron to prevent rusting *(galvanizing)*. The metal is also widely used to make alloys such as *brass, bronze* and *German silver*. Several zinc salts are employed in industry and medicine. The following compounds are of medicinal and pharmaceutical importance.

ZINC SULFATE ($ZNSO_4$)

Zinc sulfate is probably the most popular and important compound of zinc. Zinc sulfate is a colourless, transparent crystalline substance or a granular powder. It is commonly employed in medicine as an astringent, styptic and emetic. Zinc sulfate finds much use in eye washes especially in conjunctivitis.

ZINC OXIDE (ZNO)

Zinc oxide is a fine, white, amorphous powder, insoluble in water. It enjoys wide popularity either alone or in combination with other agents as a mild antiseptic, astringent and protective agent. Zinc oxide is the active constituent of two very popular external remedies: *zinc oxide ointment* and *calamine lotion*.

OTHER ZINC SALTS

Other zinc salts of some use include: *zinc chloride* (ZnCl), an antiseptic and astringent; *zinc peroxide*, an antiseptic; and *zinc undecylenate* a fungistatic agent.

Tin. Tin is a silvery-white metal which is practically inert to air and ordinary corrosive agents. For this reason tin finds its widest use as a protective material for other metals. Tinware, for example, is made by dipping sheet iron into a bath of molten tin. The common "tin can" is made in this manner. Tin is a constituent of many important alloys. *Bronze, pewter, solder, Babbitt metal* and *type metal* are familiar alloys of tin. Tin salts are of limited value and of little concern to us.

MANGANESE

Manganese is a hard, steel-like metal. Its principal use is in the making of *iron alloys* and *manganese steel.* Manganese steel is one of the hardest and toughest steels. Of the manganese compounds, *manganese dioxide* (MnO_2) and *potassium permanganate* ($KMnO_4$) are the most important. Both of these agents are extensively employed in the laboratory as oxidizing agents. In addition, potassium permanganate is employed in medicine as an antiseptic and fungistatic agent.

NICKEL

Nickel is a hard, silvery-white metal not oxidized by dry air. It is commonly used for protecting other metals. A coat of nickel is put on another metal by the process of *electroplating*. Nickel is also extensively employed in the making of valuable alloys. *German silver, nickel coin, invar, nickel steel,* and *nichrome* are popular alloys of nickel. The salts of nickel are of limited value.

CHROMIUM

Like manganese, chromium is a hard metal resembling steel. Chromium is employed in electroplating and in making certain special steel alloys such as stainless steel and chrome steel. Certain compounds of chromium are important laboratory chemicals. Of these compounds, *potassium chromate,*a yellow crystalline substance, and *potassium dichromate,*an orange crystalline substance, are the most popular. Both are excellent oxidizing agents.

GOLD

From time immemorial gold has appealed to man's economic and esthetic senses. Gold occurs almost exclusively in the free state in dust, nuggets and gold-bearing quartz. Gold is soft, heavy and bright yellow in colour; it is the most *malleable* and *ductile* of all the metals.

Chemically, it is inert to all common chemicals. *Aqua regia* (a mixture of nitric and hydrochloric acids), however, readily dissolves gold, forming gold chloride.The use of gold for jewelry and coinage is common knowledge. Since gold is so soft, it is necessary to alloy it with copper to increase its hardness.

PLATINUM

Platinum is a white lustrous metal; it is harder, heavier and more expensive than gold. Platinum has an extremely *high* melting point, and is *resistant* to all common chemical reagents. Because of these qualities, it is employed to make crucibles, wires and similar apparatus for the chemical laboratory. In a highly subdivided state, platinum is an important industrial catalyst. About 90 per cent of the world's supply of this interesting element comes from the Ural Mountains of Russia.

RADIOACTIVE METALS

The concept of radioactivity usually requires considerable exposition to impart a clear idea. For this reason, radioactivity will be thoroughly treated in a subsequent chapter. For the sake of continuity, however, let us make a few broad statements concerning radioactive metals. The story begins with the discovery of *radium* by Professor and Madame Curie. Radium is different from all the other elements heretofore known in one outstanding characteristic: The metal undergoes constant disintegration as a result of the giving-off of three invisible forms of radiation: *alpha rays, beta rays* and *gamma rays.* Although invisible, these rays have potent chemical and physiologic effects. Under carefully controlled concentrations, radium rays are useful in the treatment of cancer and tumors. At the present time, several other important radioactive metals have been produced by atomic machines such as the cyclotron. Radioactive materials are of vital importance to our present society. Later on we shall fully discuss their theoretical and practical applications.

Chapter 6

Chemical Arithmetic

An area of science *requires* and *demands* a certain degree of mathematical treatment; *this is especially true of chemistry.* Unfortunately, it seems to be well established that most people are reluctant to employ mathematics in any form. As far as the advanced disciplines of mathematics (e.g., calculus) are concerned, perhaps this reaction is justifiable.

It is not justifiable, however, when one ignores the most beautiful, applicable and simple branch of mathematics, *arithmetic.* A working knowledge of arithmetic is entirely satisfactory in disclosing the *elementary basic facts* of any study. All facts and problems presented in this chapter are resolved and solved via arithmetic.

THE METRIC SYSTEM

The metric system is the most logical system of measurement yet devised and, for this reason, is the one employed in all sciences. This system is based upon the *meter* (39.37 inches) as the unit of length; the *gram* (15.43 grains) as the unit of weight; and the *liter* (1.06 quarts) as the unit of volume. Measurements greater than or less than these units are designated decimally.

The Greek prefix *deca-*, or *deka-*, before a unit means 10; *hecto-*, 100; *kilo-*, 1000; *mega-*, 1,000,000; thus, a hectometer is 100 meters. Similarly, the Latin prefix *deci-* before a unit means 0.1; *centi-*, 0.01; *milli-*, 0.001; and *micro-*, 0.000001. The metric system is presented in the table below:

LENGTH

Millimeter (mm.)	0. 001 meter (m.)
Centimeter (cm.)	0.01 meter
Decimeter (dm.)	0.1 meter
Meter (m.)	1 meter
Dekameter (dkm.)	10 meters
Hectometer (hm.)	100 meters
Kilometer (km.)	1000 meters

WEIGHT

Milligram (mg.)	0.001 gram
Centigram (cg.)	0 01 gram
Decigram (dg.)	0 1 gram
Gram (gm.)	1.0 gram
Dekagram (dkg.)	10 grams
Hectogram (hg.)	100 grams
Kilogram (kg.)	1000 grams

VOLUME

Milliliter (ml.)	0. 001 liter (1.)
Centiliter (cl.)	0.01 liter
Deciliter (dl.)	0.1 liter
Liter (1.)	1.0 liter
Dekaliter (dkl.)	10 liters
Hectoliter (hi.)	100 liters
Kiloliter (kl.)	1000 liters

In chemistry, the *millimeter, centimeter, gram, liter, milliliter* and *cubic centimeter* are the most commonly used expressions. The *cubic centimeter* (cc.) is numerically equal to a *milliliter* (ml.). Actually, there is a very slight difference, but for all practical purposes they are used interchangeably. Obviously, it is of value to appreciate the relationship between the metric and English systems. The following table presents the important comparisons of length, weight and volume.

APPROXIMATE EQUIVALENTS

1 inch 25.4 millimeters

39.37 inches 1 meter

15 grains 1 gram

1 avoir, ounce 28 grams

1 pound 454 grams

1 fluid ounce 30 cubic centimeters or 30 milliliters

1 quart 950 cubic centimeters or 950 milliliters

HEAT

Heat is a form of energy, and is probably best understood by its influence on matter. The *Kinetic Theory* of heat states that a hot body differs from a cold body in one fundamental respect *the speed of molecular motion.* In a hot body, the molecules are moving at a tremendous rate. At a theoretical point called *absolute zero* (273° C.) all motion ceases: *This is the lowest possible temperature.* In other words, the molecular motion of a substance is proportional to the intensity of heat. From a practical point of view, however, we are interested in measuring the *intensity (temperature)* and *quantity* of heat. The concept of intensity and quantity of heat usually provokes a little confusion. These terms may be readily differentiated if we compare the quantity and intensity of heat in the flame of a match with the quantity and intensity of heat of a large body of warm water, e.g., a lake. Obviously, there is a greater amount *(quantity)* of heat in the lake than in the match flame. The intensity, though, is reversed; one will place his finger in the lake, but not in the flame.

Temperature

Temperature, or the intensity of heat, is recorded by an instrument with which we are all familiar, the *thermometer.* In this country, we are accustomed to expressing temperature measurements on the common *Fahrenheit* scale. On this scale, water freezes at 32 degrees (32°), and boils at 212°. In scientific work, the *Centigrade* scale is preferred to the Fahrenheit. On this scale, the freezing point of water is 0°, and the boiling point 100°. A comparison of the two scales will disclose that 1 scale division Centigrade equals % scale divisions Fahrenheit. A temperature on one scale may be readily converted to the corresponding temperature on the other scale by the following formula:

$$(F-32)\ 5 = 9\ C$$

Quantity of Heat

Quantity of heat is measured in *calories.* A so-called *small calorie (cal.) is the quantity of heat required to raise the temperature of 1 gm. of water 1° C.; a large Calorie (Cal.) is the quantity of heat required to raise the temperature of 1 kg. of water 1° C.* A large calorie, therefore, contains 1000 small calories. In scientific work, heat energy is usually expressed in small calories. In the English system, heat is measured in *British Thermal Units* (B.T.U.). *One B.T.U. of heat is the quantity of heat required to raise the temperature of 1 Ib. of water 1° F.* The B.T.U. is a convenient unit to express the heat value of fuels, such as coal, oil and gas.

An exact comparison between temperature (i.e., intensity) and quantity of heat is now possible by a consideration of the following problem. Compare the caloric value of 1 gm. of water at 100° C. (boiling water) with 100 gm. of water at 10° C. From our definition of calorie, we know that 1 gm. of water at 100° C. represents *100 caL;* and 100 gm. at 10° C. represents *1000 cat.* We conclude, therefore, that while the temperature of the 1 gm. of water is *10 times* the temperature of* the 100 gm. of water, it contains only *1/10* the quantity of heat.

It is often desirable to measure the calories of heat liberated during a chemical reaction. In the field of nutrition, the caloric value of various foods constitutes essential information. To measure calories, an instrument called a *calorimeter* is employed. An *accurately* weighed sample of the substance under investigation (e.g., bread, alcohol, carbon) is placed into the internal chamber, or "bomb." The contents are then ignited by means of an electrically controlled fuse. Oxygen gas is passed into the bomb prior to ignition in order to insure complete combustion. The heat liberated is thereon absorbed by an accurately measured quantity of water surrounding the bomb. Any rise in temperature is registered by the thermometer.

A motor-driven water stirrer assures an even distribution of heat. The number of calories liberated is readily calculated by multiplying the *number of grams of water times the number of degrees rise in temperature (Centigrade).* Since a weighed sample is burned, the number of calories may be expressed in terms of calories per unit weight usually *calories per gram.*

LATENT HEAT

Latent, or "hidden," heat is an interesting physical phenomenon of wide theoretical and practical importance. If a thermometer is placed in boiling water, it registers 100° C.; if it is held in the steam issuing from the water, it still registers 100° C. This would indicate that steam and water contain the same quantity of heat. That this is not true is easily demonstrated. When equal quantities of steam and boiling water are added to cold water, for example, the steam causes a greater rise in temperature (of cold water) than the boiling water does. Also, we can all testify to the fact that steam produces a much more severe burn than boiling water.

We may conclude, therefore, that a certain quantity of "hidden" heat must be added to boiling water to change it to steam. Since this added heat causes *no change in temperature,* it is an example of *latent heat,* specifically called *heat of vaporization.* All liquids in passing from the liquid to the gaseous state have heats of vaporization. The heat of vaporization is expressed as calories per gm. for water, it is 540 calories per gm.; i.e., *540 calories of heat are required to convert 1 gram of water* (at 100° C.) *to steam.* Conversely, when 1 gram of steam condenses, *540 calories are liberated.*

There is one other form of latent heat, called *heat of fusion.* The explanation is analogous to the above. Ice and ice water (i.e., water in which ice is floating) both have the same temperature (0° C.). It is easily demonstrated and verified by experience, however, that ice is "colder" than ice water. What happens is this: Additional heat *(latent heat)* must be *removed* from water at 0° C. to convert it to ice. When ice melts, this latent heat, *heat of fusion* is absorbed. For ice, *the heat of fusion is 80 calories per gm.* Every solid substance has a heat of fusion.

The true explanation of the phenomenon of latent heat is probably best given by assuming that in passing from one state to another, there is a change of molecular motion. This change involves a definite amount of heat energy. In passing from ice to water, for example, the molecular motion would increase. Since motion involves energy, heat must be added. The reverse accounts for the liberation of heat.

DENSITY AND SPECIFIC GRAVITY

Density and specific gravity are physical attributes of considerable practical importance. Every gas, liquid and solid has a definite density and specific gravity. These two terms together with their applications and calculations are described below.

Density

Density is defined as *weight per unit volume*. In chemistry, density is determined as *grams per cubic centimeter* (gm. Per cc.).

Specific Gravity

Specific gravity is a more frequently employed term than density. It is defined as the *ratio of the weight of a given liquid or solid to the weight of an equal volume of water*. A specific gravity of more than 1 means that a substance is *heavier* than water; a specific gravity of less than 1 means that a substance is *lighter* than water. The calculation is made easy by the fact that water has a density of 1 gram per cc.f.

To facilitate the determination of specific gravity of liquids, an instrument called a *hydrometer* is employed. By floating this device in a liquid, the specific gravity can be read directly at the point where a line on the scale meets the surface. The hydrometer operates on the principle that an object will float "high" in a liquid of high specific gravity, and "low" in a liquid of low specific gravity. A specially scaled hydrometer, called a *urinometer*, is used to read the specific gravity of urine. With this instrument it is possible to obtain an accurate reading to the third decimal. When using any type of hydrometer, the temperature of the liquid should be adjusted to the temperature.

THE GAS LAWS

The volume that a gas occupies depends upon the *temperature and pressure.* Unless these two factors are known it is impossible to state the volume of any gas. Since the gaseous state is commonly encountered in chemistry, we must understand and have a working knowledge of the *gas laws.* In physiology, for example, the gaseous constituents of the air and blood are an integral part of the body's chemistry. It would not be possible to comprehend their action without a knowledge of their relation to temperature and pressure. Fortunately, the gas laws which apply to pressure and temperature are easy to understand; indeed, they may be derived intuitively.

BOYLE'S LAW

In the seventeenth century the English scientist, Boyle, presented one of the most fundamental laws in chemistry. It is simple, and may be stated as follows: // *the temperature remains constant, the volume of a gas varies inversely as the pressure on it.* If, for exampje, a given gas has a volume of 100 cc. under a certain pressure, and the pressure is *doubled* (at a constant temperature), the new volume of the gas will be reduced to 50 cc.; if, on the other hand, the pressure is *halved,* the gas will *expand* to a volume of 200 cc.

CHARLES' LAW

The exact relation between the volume of a gas and its temperature was first stated by the French chemist, Charles. Like Boyle's Law, it proved to be one of the building stones of modern chemistry. Charles discovered that when a gas, under constant pressure, is heated 1° C., it expands of its volume; and, when cooled 1° C., contracts of its volume. Theoretically, then, at 273° C. it would have a volume of *zero.* It was Charles who designated the temperature of 273° C. as *absolute zero.*

Charles' Law is stated as follows: *the pressure remains constant, the volume of a gas varies directly as the absolute temperature.*

AVOGADRO'S LAW

One cannot fully appreciate the above gas law without an understanding of Avogadro's Law. Avogadro, an Italian scientist, made the brilliant deduction that *equal volumes of all gases, when measured under the same conditions of temperature and pressure, contain the same number of molecules.* The impact of Avogadro's Law upon the development of chemistry is well illustrated by the reasoning which led to the conclusion that *molecule of hydrogen contains 2 atoms, and therefore is written as* H_2. The reasoning goes like this: We know by experiment that 1 volume of hydrogen reacts with 1 volume of chlorine to produce 2 volumes of hydrogen chloride.

Let us assume that 1 liter of hydrogen and 1 liter of chlorine are reacted to produce 2 liters of hydrogen chloride. Further, let us assume that 1 liter of hydrogen contains 1000 molecules. It follows from Avogadro's Law that 1 liter of chlorine contains 1000 molecules, and that 2 liters of hydrogen chloride contain 2000 molecules. Since each of the 2000 molecules of hydrogen chloride contains 1 atom of hydrogen, there must be 2000 atoms of hydrogen. Therefore, each molecule of hydrogen gas (we started with a 1000 molecules) must contain 2 atoms of hydrogen in order to produce 2000 atoms. The above may be illustrated as follows:

1 volume + 1 volume $\rightarrow$ 2 volumes of hydrogen chloride hydrogen chlorine and so,

1000 molecules + 1000 molecules $\rightarrow$ 2000 molecules hydrogen chloride

Therefore, 1 mol of hydrogen contains 2 atoms (H_2)

By similar reasoning, it may be concluded that most of the common gaseous elements (e.g., oxygen, chlorine, nitrogen) are constituted by *diatomic* molecules.

HENRY'S LAW

Henry's Law states that the *solubility of a gas in a liquid is directly proportional lo the pressure over the liquid.* The effervescence seen when the top is removed from a bottle of pop results from the release of pressure in the bottle. Individuals who perform their work in an atmosphere of elevated pressure (e.g., *divers and caisson workers)* may develop a disease called the "bends." This condition is explained by Henry's Law: At the elevated pressures, the blood dissolves increased quantities of air. As soon as the individual returns to normal atmospheric pressure, however, this air is *released* from the blood. For this reason, great care must be exercised in adapting the body to the change of pressures; i.e., workers must *gradually* enter and leave the area of increased pressure. A sudden change from a high pressure area to atmospheric pressure allows the air to escape so rapidly that there is an actual effervescence in the blood. This causes severe pain, paralysis, and sometimes death.

MOLECULAR WEIGHT

The molecular weight (m.w.) of an element or compound is the *sum of the weights of all atoms in the molecule.* The calculation is carried out by adding the *atomic weights* of all the elements present each atomic weight being *multiplied by its subscript in the formula.*

EXPERIMENTAL DETERMINATION OF MOLECULAR WEIGHT

There is a direct mathematical relationship between the molecular weight and the freezing and boiling points of a solution. It is an interesting fact that *1 mole of a nonelectrolyte depresses the freezing point of 1000 gm. of water 1.86° C. (i.e., freezes at .86° C.),* and *raises the boiling point 0.52° C. (i.e., boils at 100.52° C.).* This laboratory fact enables the chemist to ascertain the molecular weight of a new compound. For example, if it is found that 50 gm. of a certain chemical elevate the boiling point of 1000 gm. of water 1.04° C. (i.e., boils at 101.04° C.), we may conclude that the molecular weight of this compound is 25.

The behaviour of electrolytes depends on the number of ions furnished by each molecule. Thus, 1 mole of NaCl causes a 3.72° C. depression in the freezing point, and a 1.04° C. rise in the boiling point i.e., twice the effect of a nonelectrolyte. This is due to the fact that each molecule of NaCl furnishes 2 ions. It seems, therefore, that the ion of an electrolyte and the molecule of a nonelectrolyte are equivalent in their colli-gative properties, i.e., their effects on the freezing and boiling points.

DETERMINATION OF A COMPOUND'S FORMULA

The concept of percentage composition permits the chemist to ascertain the formula of an unknown compound. When dealing with an unknown, the chemist first determines the elements present *(qualitative analysis)* and then, by using *quantitative analysis,* determines the *per cent composition of each element.* From this, the formula is determined by *dividing the percentage of each element by its atomic weight.* The resulting quotients indicate the ratio of the atoms present in each molecule of the compound.

QUANTITATIVE REACTIONS

Let us consider the following chemical reaction:

$$2NaOH + H_2SO_4 = Na_2SO_4 + 2H_2O$$

This reaction states that 2 *molecules* of sodium hydroxide react with *1 molecule* of sulfuric acid to produce *1 molecule* of sodium sulfate and 2 *molecules* of water. Since the weight of a molecule of a compound i.e., the molecular weight is constant, it stands to reason that these two chemical compounds *always react in definite amounts by weight.*

This quantitative relationship applies to all chemical reactions. It is not difficult to perceive or understand, therefore, the tremendous importance of this fundamental fact. The following examples readily illustrate the methods employed to solve problems involving chemical reactions.

Weight-Weight

In the so-called *weight-weight* problem, the weight of a reactant or product in a chemical reaction is calculated from a given weight of any other reactant or product in the reaction. In the reaction cited above between sodium hydroxide and sulfuric acid, if we know the weight of any one member of this reaction, we can calculate the three other weights. For example, from a given weight of sodium hydroxide, we can calculate the weight of sulfuric acid needed to *just* neutralize the sodium hydroxide, and the weights of the products, i.e., sodium sulfate and water. The method of calculation of *weight-weight* problems is simply illustrated below.

Example: *What quantity of sulfuric acid is needed to neutralize 10 gm. of sodium hydroxide?*

First we must write a balanced equation for the reaction, thus:

$$2NaOH + H_2SO_4 \rightarrow Na_2SO_4 + 2H_2O$$

After the equation has been correctly balanced, the given quantity of the chemical is written above its formula in the equation; an X is placed over the formula of the chemical whose weight is desired. For the above equation we have the following:

$$2NaOH + H_2SO_4 \rightarrow Na_2S0_4 + 2H_2O$$

Next, the molecular weights of the two compounds involved are determined and written under their respective formulas. In the event a coefficient is present, *the molecular weight is multiplied by this coefficient*. This is the reason why the equation must be correctly balanced.

In the same way, we can ascertain the quantities of sodium sulfate and water. It should be thoroughly understood that we are not confined to using the metric system. For example, the above problem could have read 10 oz. or 10 lbs. Since we are dealing with a proportion, however, the answer will be in the same units.

Weight-Volume or Volume-Weight

There are many reactions in which gases are either the reactants or products. As we know, when dealing with gases, volume is more significant than weight. The question arises, therefore: Can we solve reactions which involve both weight and volume? The answer is yes. By a slight alteration of the technique employed to solve weight-weight problems, we can solve weight-volume or volume-weight problems.

Volume-Volume

Some reactions involve only gases. Gas reactions are solved simply by writing the volumes in any unit above the reactants or products, and the coefficients below. In the event no coefficient is present, the number 1 is written. Gay-Lussac's Law explains why we are at liberty to write the above proportion. Even without this law, however, the above solution makes practical sense.

ARITHMETIC OF SOLUTIONS

In every area of chemistry and applied chemistry, solutions are extensively employed. It is therefore essential that we understand the principal ways in which the strength of solutions are stated. Of course, this entails the application of simple arithmetic. The following types of solutions and the necessary arithmetic are widely used in chemistry, pharmacy, medicine and nursing to mention a few.

Percentage Solutions

Percentage is the most common and popular method of stating a solution's strength or concentration. In this method, the total volume is usually taken as 100 per cent; the solute solid or liquid is taken as a percentage of the total volume. When a solid solute is dissolved in a liquid, the percentage concentration (of the solute) is calculated by *dividing the weight of the solute in gm. by the total volume in cc., and then multiplying the resulting quotient by 100.* This method of stating percentage is referred to as *weight* to *volume* (w./v.). When a solution is prepared from two or more miscible liquids, however, there are two ways of expressing the percentage of a constituent *weight to volume* or *volume to volume* (v./v.). Volume to volume is calculated by dividing volume by total volume, and multiplying the quotient by 100. To be explicit, therefore, the percentage strength of such a solution should be qualified as being w./v. or v./v.

Molar Solutions

Molar solutions are commonly prepared in the chemical laboratory. They are not as popular, however, in applied chemistry. *A 1 molar solution is a solution which contains 1 gram molecular weight of a chemical per liter of solution.* A gram molecular weight of a compound commonly referred to as a *mole* (M) is merely the molecular weight of the compound expressed in grams. Probably the principal value of employing molar solutions in chemistry arises from the fact that the term is a simple and accurate way of expressing concentration that is, to the chemist.

Normal Solutions

Normal solutions enjoy wide use in chemistry and applied areas. Normal solutions are especially valuable because of a unique quantitative relationship. In order to appreciate this relationship, it is essential that we fully grasp the meaning of the following definition: *A 1 normal (JN) solution is a solution of a compound which furnishes 1 gm. of hydrogen ions or its equivalent per liter of solution.* First of all, what is meant by the expression, "equivalent"? *An equivalent of 1 gm. of hydrogen means 1 gram atomic weight of any element which has a valency of a plus 1.* Thus, 36.5 gm. of HC1 1 gram molecular weight or mole and 58.5 gm. of NaCl 1 gram molecular weight or mole are "equivalent quantities": The 36.5 gm. of HC1 furnish 1 gm. of hydrogen ions; the 58.5 gm. of NaCl furnish 23 gm. of sodium ions (the gram atomic weight of sodium). When dissolved in sufficient water to make 1 liter of solution, each forms a 1 normal solution.

To prepare a 1 normal solution of sulfuric acid (H_2SO_4), or any other acid which contains more than 1 atom of hydrogen per molecule, *we take the appropriate fraction of the gram molecular weight.*

Thus, for dibasic acids we take *one half,* and for tribasic acids, *one third.* A 1 normal solution of sulfuric acid, therefore, contains 49 gm. (% of 98 gm.) per liter. To prepare a 1 normal solution of $MgCl_2$, or any other salt or base containing a metal whose valency is greater than 1, *we take the appropriate fraction of the gram molecular weight* e.g., %, % or y_4. A 1 normal solution of $MgCl_2$, therefore, contains 47.7 gm. (% of 95.3) per liter. In the case of salts which contain more than 1 atom of an element whose valency is greater than 1, a fraction must be employed whose *denominator is the valency of the metal times its subscript.* For example, the equivalent weight of aluminum sulfate is calculated by taking % of 1 mole of $A1_2(SO_4)3$.

Of course, normal solutions may be prepared of any desired strength 0.5 normal (0.5N), 3 normal (3N), and so on. The following examples thoroughly demonstrate how normal solutions are prepared. The careful reader has probably taken note of the fact that *a 1 normal solution and a 1 molar solution are the same for all monobasic acids and for all salts and bases with one atom of a monovalent metal.*

TITRATION

It was stated above that normality is a valuable way of expressing concentration. This is so because solutions of the same normality react with one another in equal volumes. For example, 10 cc. of IN sodium hydroxide will *exactly* neutralize 10 cc. of IN hydrochloric acid. Also, 10 cc. of 0.5N sodium hydroxide will *exactly* neutralize 5 cc. of IN hydrochloric acid. It follows, therefore, that if we know the volume and normality of one solution, we can easily calculate the volume required of some other solution of known normality. Of far greater import, however, is the fact that it is possible to determine the normality of a solution. This relationship between normal solutions is expressed by the following simple formula:

N X cc. = NX cc.

(solution A) (solution *E)*

Example: *What is the normality of a solution of silver nitrate which requires 30 cc. of 2N sodium chloiide to react with 60 cc. of solution?*

Although not required for solution, the equation for the reaction is:

$$AgNO_3 + NaCl = NaN0_3 + AgCl$$

In other words, 30 cc. or 2N sodium chloride *exactly* react with 60 cc. of xN silver nitrate to form sodium nitrate and insoluble silver chloride. The normality of the $AgNO_3$ solution is calculated as follows:

$AgNO_3$ NaCl

N X cc. = N X cc.

N X 60 = 2 X 30

N60 = 60

N = 1

How are problems such as the one cited above carried out? Specifically, how is it possible to determine the point at which one solution *exactly* reacts with another solution? More specifically, how do we know that 30 cc. of 2N sodium chloride exactly react with 60 cc. of silver nitrate solution? The answer to these questions is supplied by the word *thration. Titration is defined as the process of determining the normality of an unknown solution by using a solution of known normality.* The silver nitrate-sodium chloride reaction is carried out as follows: Burette A is filled with silver nitrate solution, and burette B with the sodium chloride solution.

The stopcock of burette A is then opened and exactly 60 cc. of solution are delivered to the beaker. The beaker is then placed under burette B and *titrated* with the sodium chloride solution; i.e., the solution is allowed to enter the solution in the beaker *drop* by *drop.* As the sodium chloride solution enters the silver nitrate solution, a white precipitate of AgCl forms immediately. When carefully conducted, a point the so-called *end point* will be reached at which no further precipitate will form. At the end point, *just enough* of the solution in the burette has been added. The volume of sodium chloride solution added can then be read directly from the burette.

For reactions which do not form precipitates, the end point is ascertained by a so-called *indicator.* An indicator usually a dye will change colour at the *end point* of the reaction. When titrating a base with an acid, for example, a drop of phenolphthalein solution is added to the base; this imparts a red colour. At the point of neutrality the end point the red colour disappears, and titration is stopped. Other indicators, such as litmus and methyl orange, may be similarly employed.

Titration has wide application in *quantitative analysis* the branch of chemistry which is concerned with ascertaining the concentration of an element or compound. The following example illustrates its value in industry. Vinegar as sold on the market is required to have a concentration of 4 to 5 per cent acetic acid.

Obviously, therefore, all batches of vinegar sold by a manufacturer must be subjected to a quantitative analysis before being bottled. This is carried out by titrating a sample of the vinegar with a base such as sodium hydroxide. The normality of the vinegar is then calculated as above. From the normality we can calculate the amount of acetic acid and the percentage concentration. Suppose, for example, that a manufacturer has a batch of vinegar with a normality of 0.5. Knowing this, he can find the percentage of acetic acid.

Chapter 7

Understanding the Colloids

The subject of colloids demands a special chapter and special consideration. Heretofore, we were dealing primarily with the chemical properties of the elements and their compounds; in contrast, colloidal properties are essentially physical in nature. *A colloid may be defined as a solid, liquid or gas dispersed (subdivided) to such a degree that the size of the dispersed particles becomes an important factor in determining its properties.* It is hoped that a careful study of this chapter will bring full understanding and appreciation of this definition. The vast importance of colloidal chemistry especially to the biologic sciences will be amply disclosed throughout the chapter.

THE COLLOIDAL STATE

When sand is shaken with water and the mixture is allowed to stand, the sand soon settles to the bottom. Such a mixture is a true *suspension.* If, on the other hand, we add sugar to water, the molecules become separated from the crystals and are distributed uniformly throughout the solvent. As we know, such a mixture is a true *solution.* Now let us grind some clay into small particles, and stir these thoroughly into a beaker of water. When such a mixture is allowed to stand, the greater amount of the clay owing to its weight and insolubility settles to the bottom. If we filter this mixture through filter paper, the clay which settled out will remain on the paper. The liquid which passes through, however, is not clear but *turbid.* This turbidity is due to the particles of clay which have passed through the pores of the filter paper. Of fundamental importance is the fact that these clay particles do not settle out, but remain suspended in the liquid. These particles of clay are said to be in the *colloidal* state.

The size of an individual colloidal particle is so small that it is usually measured in *millimicrons.* One millimicron, is equal to one-millionth of a millimeter. A colloidal particle is usually somewhere between 1 nv (0.000001 mm.) and 100 nv (0.0001 mm.); i.e., it is smaller than a suspension particle, but larger than a solution particle.

COLLOIDAL SYSTEMS

The above example of colloidal clay in water represents a so-called *colloidal system.* A colloidal system may be defined as a two-phase heterogeneous mixture i.e., one state of matter dispersed in another state of matter in which the dispersed particles do not settle out. The phase which is suspended is called the *dispersed phase* or *internal phase;* the phase which holds or contains the dispersed phase is called the *dispersion medium* or *external phase.* In the example cited, clay represents the dispersed phase, and water the dispersion medium. The dispersed phase may be a solid, liquid or gas. Likewise, the dispersion medium may be a solid, liquid or gas. Theoretically, therefore, there are nine possible colloidal systems. Actually, however, there are only eight systems, since gases are miscible in all proportions; i.e., gases form true solutions. These eight colloidal systems are presented in the table below:

CLASSIFICATION OF COLLOIDAL SYSTEMS

Dispersed Phase	Dispemon Medium	Name
Solid	Liquid	Sol
Solid	Gas	Solid aerosol
Liquid	Solid	Solid emulsion
Liquid	Liquid	Emulsion
Liquid	Gas	Liquid aerosol
Gas	Solid	Solid foam
Gas	I jiquid	Foam
Gas	Gas	

COLLOIDAL PROPERTIES

Before we study the individual colloidal systems, let us first direct our attention to the unique properties common to all colloids.

Tyndall Effect. Probably the most striking property of colloidal systems is the so-called *Tyndall effect*. This phenomenon is observed when a beam of light is passed through any colloid: *The path of light may be clearly seen.* The effect is similar to that observed when a beam of sunlight passes through a darkened room. A true solution, on the other hand, does not show the path of light because the molecules of solute are too small to *reflect light*. The following experiment vividly demonstrates the Tyndall effect. Cigarette smoke is blown into a large jar. The room is then darkened, and a small pencil of light is directed through the smoke; *the smoke particles clearly outline the path of light*. The Tyndall effect is therefore a valuable aid in determining if a certain mixture is a true solution or a colloid.

BROWNIAN MOVEMENT

An instrument called the *ultramicroscope* greatly magnifies the Tyndall effect. The essential parts of an ultra-microscope are a small cubical cell of quartz to hold the liquid, a powerful source of light to send a pencil of rays through the liquid, and a microscope mounted at right angles to the beam of light and against a dark background. Colloidal particles are seen as bright specks of light in *rapid zigzag* motion. This zigzag motion or *brownian movement* (first observed by the English botanist, Brown) remained unexplained for fifty years. It is now known to be a necessary consequence of the high speed molecules of the solution colliding with the colloidal particles.

ELECTRICAL CHARGE OF COLLOIDAL PARTICLES

Colloidal systems are fairly good conductors of electricity. Unlike solutions of electrolytes, however, which furnish both positive and negative ions, colloidal systems are either *all* positive or *all* negative. It appears that a colloidal particle gets its charge by the adsorption of some simple ion already present in the solution. If the dispersion medium happens to be water, the charges are probably furnished by the H^+ ions or OH^- ions. The electrical character of a colloidal system may be demonstrated by the apparatus. When a direct current is passed through a colloid in the U-tube, all the particles migrate toward the anode or cathode, depending upon their charge. In the case of a coloured colloid, such as ferric oxide dispersed in water, the colour will disappear from one of the arms of the tube. This migration of colloidal particles toward an electrode under the influence of an electric current is called *cataphoresis*. Cataphoresis is of paramount importance in colloidal chemistry.

The fact that colloidal particles in any one colloidal system have the same charge is really the backbone of the colloidal state. *These charges serve to keep the colloidal particles apart by repulsion.* This theory is strongly supported by experimental evidence. If we add an electrolyte to a colloidal system, *precipitation* results. The large number of both kinds of ions introduced by the electrolyte neutralizes the charged colloidal particles. These neutral particles then grow into larger and larger particles until a precipitate forms. In this manner, the introduction of an electrolyte destroys a colloid by precipitating the dispersed phase. As a rule bivalent ions (e.g., Ca^{++}) are more effective than univalent ions (e.g., Na^+), and trivalent ions, still more so. Also, if we mix two colloids of opposite sign, the particles of one will attract the particles of the other; the two then form a neutral clump which grows larger and larger with subsequent precipitation.

ADSORPTION

First, it is appropriate to point out the distinction between *adsorption* and *absorption*. By adsorption we mean the adherence of molecules or ions of any kind to the surface of a solid or liquid, while by absorption we mean the penetration of the molecules or ions into the interior of the solid or liquid. Actually, both processes often take place simultaneously.

Colloids adsorb substances to a high degree. This is due to the immense surface area of colloidal particles. For example, a cube having a length of 1 centimeter on each side has a total surface area of 6 square centimeters. If this cube is divided into cubes which measure 0.0001 mm. on each edge i.e., subdivided into the colloidal state the total surface would be 600,000 square centimeters. It is quite evident that these "colloidal cubes" are able to adsorb more molecules of another substance than the larger cube can.

The property of adsorption is often taken advantage of in the laboratory and industry. Charcoal and boneblack (a form of charcoal from bones) due to their vast surface area per gram are excellent adsorbing agents. Charcoal is usually employed to remove odorous gases; bone-black is used to remove undesirable colouring matter.

COLOUR

Many colloidal systems are highly coloured. Colloidal colour is explained by the fact that the minute particles *absorb certain wavelengths of light and reflect others.* The specific colour, however, depends primarily upon the size of the particle rather than its character. For this reason the colour gives little clue as to what substance is present. Thus, colloidal gold may be red, blue, green or violet, according to the size and uniformity of the particles.

It takes only a very small amount of colloidal substance to impart an intense colour. Just a few milligrams of gold, for example, will colour a liter of water. Red glass is manufactured by adding just a trace of selenium to ordinary glass. The blue of the sky and of the ocean is explained on a basis of colloidal dust in the air and water. It is thought that the blue of the eyes is due not to pigment, but to the presence of colloidal particles in the iris which scatter blue light and transmit red. The brilliant colours of some of the feathers of many birds are due to the presence of colloidal air bubbles.

DIALYSIS

When a colloidal solution is filtered, no solid is retained on the paper; i.e., the particles unimpeded through the pores of the paper just like a solution. We find, however, that if the filter paper is replaced by a parchment or cellophane membrane, the colloidal particles are prevented from passing through. Solution particles (molecules and ions), on the other hand, pass through with ease. From a standpoint of particle size, therefore, there are two types of liquid mixtures: *mixtures whose particles do not pass through parchment or animal membranes,* and *mixtures whose particles do pass through.* The latter type of mixture, as we know, constitutes a true solution. The term *crystalloid* is applied to all substances which form true solutions. The terms *colloid* and *crystalloid* are therefore chemical antonyms.

Often times it is desirable and essential to separate substances in solution from substances in the colloidal state, i.e., to separate crystalloids from colloids. This can be easily accomplished by placing the liquid mixture in a parchment bag, and suspending the bag in water. After a certain period of time the crystalloid particles diffuse through the membrane; the colloidal particles do not. This process of separating the crystalloid state from the colloid state is called *dialysis.*

Dialysis has many applications, and explains many facts. In the extraction of medicinal products from plant and animal tissues, for example, dialysis affords the manufacturer with a unique and efficacious way to separate the active constituent from the colloidal contaminants. The membrane which surrounds a living cell has selective properties; *it prevents the loss of protoplasm (a colloid) and permits the entrance of nutritional crystalloids.*

METHODS OF PREPARING COLLOIDS

There are many methods and ways of preparing colloidal systems, but all of these may be classified into two groups *dispersion and condensation.*

Dispersion

Dispersion consists in subdividing larger aggregates into particles of colloidal size. We start with an insoluble substance and by some means reduce it to particles of colloidal size, and then mix these particles with the desired dispersion medium.

Grinding and agitation are common ways of reducing particle size; grinding is applied to solids, and agitation to liquids. At the commercial level, a so-called *colloid mill* is used to render a substance to the colloidal state. The mill consists of two rapidly revolving surfaces, about 0.002 inch apart, between which particles in suspension are drawn. The colloid mill is used to prepare colloidal substances such as paints. inks, ointments and emulsions.

In many cases dispersion is carried out by treating a finely divided precipitate with some chemical. The chemical by some means not chemical reduces the precipitate to colloidal magnitude. Dilute sodium hydroxide, for example, causes clay particles to become colloidal. This action is called *peptization,* and the chemical employed is said to *peptize* the precipitate.

Colloidal systems of metals (e.g., gold, platinum, silver) dispersed in a liquid may be prepared by a unique method of dispersion: An arc is passed between two wires of the metal held under the liquid.The arc causes colloidal particles of the metal to be separated from the wire, and spread throughout the solution.

Condensation

Condensation consists in the building up of colloidal particles from smaller particles i.e., from molecules or ions. This is usually done by forming a colloidal precipitate via chemical action between two solutions. When an acid is added to a dilute solution of sodium thiosulfate, for example, colloidal sulfur is formed, and a colloidal system of sulfur dispersed in water results. Many colloidal systems may be prepared in this manner. Since the formation of a colloid is probably the first step in precipitation, chemical condensation must be carefully controlled and stopped before the particles reach the upper limits of colloidal dimensions.

Condensation can also be accomplished by the addition of a liquid to a solution of a substance which is extremely insoluble in the liquid added. For example, if water is added drop by drop to a solution of benzoin in alcohol, benzoin is released from solution. When carefully carried out, i.e., if just the right amount of water is added, a colloidal system of benzoin dispersed in an alcohol-water medium results.

SOLS

Sols are colloidal systems of solids dispersed in solids, liquids or gases. There are three types of sols according to our classification: *sols, solid sols and solid aerosols.* These three types are described below.

Sols. A sol is a colloidal system of a solid dispersed in a liquid. Some are natural, and some may be prepared artificially. Sols are classified as *emulsoids* or *suspensoids.*

EMULSOIDS

Nearly all naturally occurring colloidal systems are emulsoids. Emulsoid sols are formed from *lyophilic* substances. The term lyophilic is applied to a substance which upon dispersion shows an attraction or affinity for the dispersion medium, swelling as it passes into the colloidal condition. For this reason, emulsoids are much more viscous than the dispersion medium alone, and when concentrated often form *gels,* i.e., a jelly-like colloid. The higher viscosity is associated with the swelling of the particles as they become more and more "solvated," increasing the volume of the dispersed phase. Emulsoids are generally quite stable; they are especially resistant to electrolyte precipitation.

Emulsoids are usually prepared with ease, due to the lyophilic nature of the dispersed phase. Gelatin, glue, starch, proteins and gums are examples of lyophilic substances. A sol of starch or gelatin, for example, is prepared by merely adding warm water. Actually, this is a process of peptization, where the water brings about the dispersion of the particles.

The fact that protoplasm is best described as a jelly-like substance is indeed of singular importance. There is reason to believe that most of the organic substances within plant and animal tissues are of a colloidal character, and display the properties described in this chapter. Therefore, when dealing with biologic chemistry, we must appreciate the fact that we are not dealing with pure solutions; indeed, we must realize that nature is colloidal.

SUSPENSOIDS

A suspensoid sol is one in which the suspended particles do not tend to combine with the dispersion medium. Suspensoids are prepared from *lyophobic* substances, i.e., substances which do not have an attraction or affinity for the dispersion medium. In contrast to emulsoids, suspensoids are not viscous, and they are not as stable, being susceptible to electrolyte precipitation.

Suspensoids rarely occur in nature. They may be prepared by dispersion or condensation methods. The dispersed phase is almost always an inorganic substance. Clay, sulfur and gold sols are examples of suspensoids.

The formation of deltas at the mouths of rivers is explained by colloidal chemistry. As the river flows toward the sea, clay becomes dispersed in the waterÿÿi.e., a suspensoid is formed. When the sea water comes into contact with the clav particles, the positively charged sodium and magnesium ions contained in the sea water neutialize the negatively charged clay particles. As a result, the particles coagulate into larger masses and settle at the mouth of the river. This deposition of clay at the bottom forms the delta.

SOLID AEROSOLS

A solid aerosol is a colloidal system of a solid dispersed in a gas. Solid aerosols are commonly referred to as *smokes.* Smoke is produced from the incomplete combustion of carbonaceous materials such as coal, oil and tobacco; it also results from the condensation of vapors to the solid state. Certain dry gases combine chemically to form a smoke. For example, if a drop of hydrochloric acid is brought close to a drop of ammonium hydroxide, a dense white smoke of ammonium chloride is formed.

Enormous amounts of money are lost at the industrial level from the loss of valuable materials which escape via the smoke stack. Precious metals, zinc and potassium compounds are but a few examples. Ordinary carbon smoke causes great loss to the individual but a profit to the laundry.

An ingenious application of colloidal chemistry by Cottrell has presented industry with a solution to its smoke problems. The so-called *Cottrell Process* has proven to be a panacea for the health of those living in industrial areas, as well as the pocketbooks of industry. In this process, smoke is passed between two plates (usually installed at the bottom of the stack) maintained at a high voltage. The charged particles of the smoke arc drawn to the plate of opposite charge and neutralized. In this manner a dust forms on the plates which quickly settles.

SOLID SOLS

A solid sol is a colloidal system of a solid dispersed in another solid. The best examples of solid sols are coloured glass and precious stones such as ruby, sapphire and amethyst. Red glass is made by adding a little selenium to ordinary glass; ruby glass contains gold dispersed in glass. The colour of the iris of the eye and other anatomic structures (e.g., skin) is best explained on a colloidal basis. As pointed out above, the colour results from the absorption and reflection of light by the dispersed particles.

EMULSIONS

Emulsions are colloidal systems of liquids dispersed in liquids. When two immiscible liquids, such as oil and water, are shaken, a milky fluid results; this is an emulsion. The agitation has dispersed the oil into globules of colloidal dimensions. Such an emulsion is referred to as oil in water emulsion. If this emulsion is allowed to stand, it will soon separate into the original liquids from which it was made. Emulsions, however, can be made stable by the addition of a third substance called an *emulsifying agent.* An emulsifying agent acts it seems by forming a film over the surface of the dispersed globules. If to our oil in water emulsion mentioned above we add just a little soap, a *stable* emulsion results. The film evidently prevents the globules from running together to form larger globules.

Emulsions and emulsifying agents are of considerable theoretical and practical importance. The fats and oils contained in protoplasm are undoubtedly in the form of emulsions. Milk is an emulsion of butter fat dispersed in water, with *casein* as the emulsifying agent. The souring of milk is explained thus: Certain bacteria present in the milk act upon lactose to form lactic acid; the acid coagulates the casein, and the emulsion is thereby destroyed. Mayonnaise is a highly concentrated emulsion of. vegetable oils dispersed in water, with egg yolk as the emulsifying agent. Countless products on the market are emulsions. Cod liver oil and mineral oil emulsions are prepared by the pharmacist in order to increase palatability. Cresol, a commonly used disinfectant, would be of little value if it were not in the form of an emulsion. Lysol is a trade name for a soap emulsion of cresol.

During the process of digestion, emulsification plays an important role. In order for fats and oils to be properly digested by the enzymes within the intestine, they must be first emulsified. The fats and oils in the colloidal state i.e., emulsified are readily acted upon; *this is due to the greater surface area of the colloidal state.* The bile salts present in bile serve as the emulsifying agents. Any obstruction or dysfunction which interferes with the emulsification of fats and oils in the diet (e.g., jaundice, gallstones) is a nutritional insult.

SURFACE ACTIVE AGENTS

All emulsifying agents have one characteristic in common: *They reduce surface tension.* By surface tension of a liquid we mean in simple but fairly accurate language its tendency to assume a shape which gives the smallest surface area possible. For example, if a drop of alcohol and a drop of water are placed on a glass plate, there is considerable difference between the areas covered by the drops; alcohol, having *less* surface tension than water, covers a greater area. Therefore, when an emulsifying agent is added to a liquid, the liquid is more able to *spread* and *penetrate.*

In the last few years chemicals have been synthesized which possess tremendous powers of emulsification; i.e., they drastically lower surface tension. Such agents commonly referred to as *surface active agents* have revolutionized the cleaning industry. *Tide, Vel, Drejt, Fab, Cheer* and similar "soap" products contain surface active agents. Such products are commonly referred to as "detergents."

Certain surface active chemicals have proven to be potent bacterio-static and bactericidal agents. Their action is ascribed to their powers to penetrate pathogenic organisms. *Zephiran,* a popular surface active antiseptic and germicide, is capable of inhibiting bacterial metabolism at a dilution as great as 1 to 30,000. Unlike conventional antiseptics and disinfectants, agents such as Zephiran are devoid of irritating properties.

OTHER COLLOIDAL SYSTEMS

The following colloidal systems deserve only brief mention.

Liquid Aerosols

Liquid aerosols are colloidal systems of liquids dispersed in gases. Liquid aerosols include atmospheric fog, clouds and mists. Preparations referred to as "aerosols" are commonly used as disinfectants, deodorants and insecticides. A DDT aerosol, for example, is formed by allowing air pressure to force a solution of DDT through a very small valve; in this way, particles of the solution enter the air in a colloidal state. Aerosols have proved to be an effective and economical way of applying such agents.

Solid Emulsions

Solid emulsions are colloidal systems of liquids dispersed in solids. Water in paraffin wax and minerals with liquid inclusions (e.g., opal, pearl) are examples of this type of system.

Foams

Foams are colloidal systems of gases dispersed in liquids. Certain agents enhance foam formation. Licorice, for example, when added to a solution in which carbon dioxide gas is being evolved causes a copious heavy foam to form. This is the reason why licorice extract is used in the Foamite type of fire extinguisher. Usually foams are formed by vigourous agitation of a liquid. Anyone who has beaten egg whites has witnessed the beautiful foam which results.

Solid Foams

Solid foams are colloidal systems of gases dispersed in solid. Gaseous inclusions in minerals (pumice), lavas, hydrogen in iron and adsorbed gases are examples.

Chapter 8

Organic Chemistry

There are two types of compounds which constitute the earth: *organic* and *inorganic*. Organic compounds include all compounds which contain carbon; all other compounds are referred to as inorganic. In general terms, therefore, there are two great branches of chemistry: *organic, which is defined as the study of carbon and its compounds; and inorganic, which is defined as the study of all the other elements and their compounds.*

What is there about the element carbon that places it in a category by itself? Actually, carbon is the most unique of all the elements: *Carbon atoms have the ability to combine with one another an endless number of times, and in an endless number of ways.* This fact explains the huge number of carbon compounds found in the tissues of plants and animals, and in the products derived there from (e.g., petroleum, coal tar, wood).

In addition to their number, organic substances are also characterized by the extremely complex nature of their molecules; molecular weights in the vicinity of hundreds of thousands are common in compounds such as proteins and related substances. It is obvious, therefore, that the study of organic chemistry must be approached and handled in a manner slightly different from our previous studies. In order to effectively grasp the principles of organic chemistry, we must learn to *classify* and *generalize;* i.e., we must apply techniques similar to those employed in biology. The present chapter sets forth a synopsis of the cardinal concepts in a manner commensurate with this view.

Before we embark upon our journey throughout the heavens of carbon atoms, it will be instructive to compare our present definition of organic chemistry with the definition of the preceding century. At that time, organic chemistry was defined as *the chemistry of living things and their products.* From what has been said, this was a reasonable and logical definition. In the year 1828, however, the German chemist Wohler completely changed the picture by an experiment which was as outstanding in its simplicity as in its farflung implications. Wohler discovered that the chemical compound *ammonium cyanate* could he converted by heat into *urea* a substance theretofore produced only in the tissues of animals and excreted in urine. *Thus, Wohler demonstrated that it was possible to synthesize a natural product.*

Other syntheses followed; today the organic chemist seems to be able to synthesize any compound of nature. Moreover, he can prepare substances foreign and, in many cases, superior to those of natural origin. Plastics, rubber, dyes and drugs are but a few examples of his ability. Organic chemistry is best defined, then, as simply the *study of carbon and its compounds.*

The importance of organic chemistry may be easily perceived from the above statements. There is not a single area of the biologic sciences which does not demand some knowledge of the basic principles. In the following chapter on biochemistry, i.e., the chemistry of life processes, an understanding of organic chemistry is a *sine qua non.* We shall do well, then, to assimilate thoroughly the concepts of the present chapter.

Nonelectrolytes. *In contrast to inorganic substances, organic compounds are nonelectrolytes.* This is to be expected since organic molecules are constituted by *covalent atoms;* i.e., no ions are available.

In general, a substance which does not conduct an electric current is an organic substance. This test, plus the fact that *organic compounds char or burn upon heating,* enable us to identify swiftly organic substances.In general, organic compounds react at a much *slower rate* than inorganic compounds; this is easily explained by their nonionic nature. It will be recalled that compounds which furnish ions react immediately. When a solution of salt is added to a solution of silver nitrate, for example, a copious white precipitate forms instantly. On the other hand, there are many organic reactions which require several hours to react completely. The covalent nature of the organic molecule, then, is a basic chemical difference.

STRUCTURAL FORMULAS

The properties of an organic compound depend upon the *arrangement* of its atoms as well as the number and kinds of atoms in the molecule. This is epitomized by the striking fact that we often find two or more compounds to which the same ordinary, or *empirical,* formula must be assigned. The organic chemist has conclusive evidence that these different compounds are a result of different arrangements of the atoms in the molecules. For this reason and many others the *structural formula* is employed in organic chemistry.

For the sake of time and convenience, structural formulas are usually abbreviated or condensed. Any such formula is acceptable so long as the complete structure is correctly indicated.

The ability to write and interpret structural formulas is essential. The above facts in conjunction with others contained in this chapter will easily permit us to master the art.

ISOMERISM

Let us now direct our attention to those compounds which have the *same* empirical formula, but *different* properties. As stated before, this situation is common in organic chemistry. The term *Isomerism* is applied to this condition; the compounds are referred to as *isomers.* Isomerism can be explained only on a basis of a *difference* in molecular structure. Let us look a little closer and see if we can deduce a little more information about this phenomenon.

OPTICAL ISOMERISM

A peculiar type of isomerism is shown by many organic substances. Lactic acid will serve as a typical example. All three varieties of this acid possess the *same* physical and chemical properties they *differ* only in regard to their effect upon *polarized light;* hence, they are said to be *optically active.* When polarized light is passed through each of the three acids the following happens: One kind turns the plane of polarized light to the *right* (this acid is called *dextro* or d-lactic acid); the other acid turns the plane of light to the *left* (this is *levo* or /-lactic acid); and the third has no effect upon polarize J light it is called lactic acid.

When ordinary light is passed through a glass plate of special construction, such as a tourmaline plate, it emerges changed in this respect: The electromagnetic vibrations take place in only *one* plane, as opposed to ordinary light which vibrates in all planes; such light is said to be *polarized.* Now, when polarized light falls on another plate of the same construction as the first, it will completely pass through *only* if the second plate is in a certain position specifically, only when the crystals of the two plates run in the same direction. Other positions will allow varying degrees of light to pass all the way to a complete obstruction.

Let us place two such plates at the ends of a glass tube a few inches long; further, let us make one stationary (this plate is called the *polarizer),* and the other so that it rotates (this plate is called the *analyzer).* Now let us view a source of light with the eye at the analyzer, and rotate the analyzer so that maximum light is seen. With the analyzer in this position, an *optically active* compound (in solution if a solid) is placed into the tube. If the variety added is *d,* the light as viewed with the eye at the analyzer is *not* as bright as before; this is because the plane of light leaving the polarizer has been *turned* by the solution, with the result that it does not hit the analyzer at the proper angle for complete transfer.

By rotating the analyzer, however, we can once again restore maximum brightness. By convention, if the adjustment is made by rotating to the *right,* the compound is in the *dextro* or *d* form; if the analyzer must be turned to the *left,* the compound is called the *levo* or form. The *dl* form of a compound *does not* change the plane of light. Actually, the *dl* form is really a mixture of the form and *d* form.

In practice it is desirable to know the exact or so-called *specific rotation* of an optically active compound. The specific rotation of a compound is defined as the number of degrees the analyzer must be rotated for a solution containing 1 gm. of the compound per cc. in a polarimeter tube 1 dm. long at a specific temperature and at a specific intensity of light.

HYDROCARBONS

All organic compounds may be considered derivatives of the great class known as *hydrocarbons. A hydrocarbon is an organic compound which contains only carbon and hydrogen.* Hydrocarbons range in complexity from a few atoms per molecule to compounds whose molecules contain in the neighbourhood of a hundred atoms. Hydrocarbons occur naturally in petroleum and coal tar. Actually, petroleum is nothing more than a mixture of hydrocarbons. In this section we shall discuss the various classes of hydrocarbons and the important members thereof.

There are two major categories of hydrocarbon compounds; i.e., any one hydrocarbon has chemical and physical properties which place it into one of two groups of similar substances. Thus, out of their vast number the organic chemist has demonstrated an unyielding devotion to obey certain rules and laws. It has been deduced from extensive studies that these two major categories are a result of the way in which the carbon atoms are joined in the hydrocarbon molecule. Conclusive evidence indicates that they are joined in *chains* with open ends, or in *rings* without open ends. Hydrocarbons of the chain type are referred to as *aliphatic* hydrocarbons; hydrocarbons of the ring type are referred to as *cyclic* hydrocarbons. The cyclic hydrocarbons, in turn, are of two classes: *aromatic hydrocarbons* and *alley die hydrocarbons.*

PETROLEUM

Petroleum is the vast natural source of the aliphatic hydrocarbons. It is a dark-coloured oily liquid composed of gaseous and solid hydrocarbons dissolved in liquid hydrocarbons. By a process called *fractional distillation* the many vital constituents are separated. Fractional distillation is the process of separating a mixture into parts, or *fractions,* by distilling the mixture and collecting the parts of the distillate separately. This is possible by virtue of the different boiling points of the substances in the mixture. In addition to the pure hydrocarbons presented in the tables, petroleum yields an endless number of useful products such as *gasoline, kerosene, mineral oil, Vaseline, paraffin* and *lubricating* oils. These products are not single compounds, but *mixtures* of hydrocarbons which appear in certain fractions. The world production of petroleum is now over 2 billion barrels annually.

METHANE SERIES

The hydrocarbons of this series are often termed *paraffins* (paraffin is a mixture of solid hydrocarbons of this series). The table below gives the names, formulas and boiling points of the first ten members. Others are known as far up as 28 carbon atoms; the first 10, however, are the most important.

METHANE SERIES

	Formula	***Boiling point*** **(°C)**
Methane	CH_4	-161
Ethane	C_2H_6	- 88
Propane	C_3H_8	
Butane	C_4H_{10}	- 44- 6
Pentanc	C_5H_{12}	36
Hexane	C_6H_{14}	69
Heptane	C_7H_{16}	
Octane	C_8H_{18}	98124
Nonane	C_9H_2O	150
Deeane	$C_{10}H_{22}$	174

A glance at the table will disclose the singular fact that each member of the series differs from the one preceding it by the group of atoms, CH_2. Such a series of hydrocarbons is called a *homologous series*. Further, the formula for any member of the series is given by the general formula, $C_nH_{2n}+_2$, in which "n" is the number of carbon atoms. In general, the members of a homologous series are similar in their chemical characteristics. Their physical properties show a regular gradation; i.e., their physical state (gas, liquid or solid) is a function of their molecular weight.

PROPERTIES

The hydrocarbons of this series are all *colourless* compounds. The first four members methane, ethane, propane and butane are gases; those containing 5 to 16 carbon atoms are liquids; and those containing more than 16 carbon atoms are solids. *All members of this series are insoluble in water. Chemically, they are rather inert; strong acids and bases have little or no effect upon them. All of them burn, however, to form carbon dioxide and water.*

METHANE

The general properties of the paraffins are characterized by methane. Methane gas also known as *marsh gas* and *fire damp* is the simplest compound of carbon. Methane constitutes from 90 to 95 per cent of natural gas. It is formed in marshes by the decay of organic matter under the water; bubbles of the gas are often seen rising from the bottom of stagnant pools. In the laboratory the gas is prepared by heating a mixture of sodium acetate, sodium hydroxide and calcium hydroxide.

Methane is a colourless, odorless gas, about one-half times as heavy as air. It is only slightly soluble in water. Chemically, methane is rather inert; it fails to react with most substances. In air, however, it burns with a very hot, pale blue flame, forming carbon dioxide and water:

$$CH_4 + 20_2 = C0_2 + 2H_2O$$

In the presence of sunlight, methane reacts with chlorine and bromine to form a series of products. These derivatives will be discussed later.

ETHYLENE SERIES

Ethylene is a hydrocarbon which has the formula C_2H_4. If we attempt to write a structural formula for this compound, we find that the laws of valency seem to be violated.

Ethylene is the first member of the so-called *ethylene series* of hydrocarbons. Each member, like ethylene, is characterized by the Ethylene is the first member of the so-called *ethylene series* of hydrocarbons. Each member, like ethylene, is characterized by the possession of a *double bond.* The general formula for hydrocarbons of this series is C_nH_{2n}.

PROPERTIES

Like the methane series, the lower members are gases and the higher members are liquids or solids. Chemically., however, there is an important difference between the two series. In contrast to the paraffins, members of the ethylene series are *decidedly reactive.* This can be explained only on a basis of the *double bond.* When given the opportunity, it appears that the carbon atoms which share a double bond prefer to dissolve the relationship; i.e., a double bond imparts an *unstable* nature to a molecule. The halogens, for example, swiftly break the bond and *add* on to the molecule.

A hydrocarbon or any other organic compound which possesses a double bond is said to be *unsaturated;* i.e., it is capable of adding on atoms. The compound formed as a result of the addition is termed an *addition product.* Conversely, compounds which do not contain a double bond are said to be *saturated.* Saturated compounds react by *substitution;* i.e., atoms are exchanged for hydrogen. The concept of the double bond is of vital import.

ETHYLENE

The general properties of the series are characterized by ethylene. Ethylene is a colourless gas which can be condensed to a liquid. It burns with a luminous flame, forming carbon dioxide and water. Us most characteristic property as indicated above is its power to unite (by addition) with other substances, particularly with the halogens.

Ethylene may be prepared by a number of methods. Of special interest is its preparation by the removal of the elements of water from ethyl alcohol (C_2H_5OH). This can be accomplished by the use of a suitable dehydrating agent such as sulfuric acid. Ethylene gas and its derivatives are very useful substances. Of singular interest is the fact that ethylene is a potent anesthetic. Its highly explosive nature, however, presents a serious disadvantage.

ACETYLENE SERIES

Acetylene gas is by far the most important member of a series. Members of this series are unique in that they possess a *triple bond.* The properties of acetylene are illustrative of its homologs. Physically, acetylene is colourless, possesses a garlic-like odor, and is sparingly soluble in water. Chemically, acetylene is marked by extreme activity. Thus, the *triple bond* imparts a more unstable character to a molecule than does the double bond.

Commercially, acetylene has been used for cutting and welding metals, and as a source of illumination. Acetylene is also used in the manufacture of a number of organic compounds of industrial importance. Thus, it is possible to start with *lime* (calcium oxide) and *coke* (carbon) simple and cheap raw materials and synthesize complex substances.

NOMENCLATURE

The ability to name an organic compound properly is probably the most important practical aspect of organic chemistry. In the early years of the study, naming was haphazard; indeed, the chemists themselves were often confused. As the number of compounds grew by leaps and bounds, it became quite evident that an accurate and explicit system of nomenclature was essential. Today almost all organic chemists throughout the world have agreed upon a basic system. It is evident, of course, that the "old" and the "new" are always coexistent. The name *chlorofonn,* for example, will probably continue in preference to the more accurate name, *trichloromethane.* In general, however, confusion has given way to order. It is hoped that the student will develop in an elementary way a "feel" for naming organic compounds. The aliphatic hydrocarbons are named as follows:

METHANE SERIES

The name for any member of the *methane* series has the suffix *-ane*. With the exception of the first four members, the first part of the name indicates the number of carbon atoms in the molecule. When one hydrogen atom is removed from any member of the methane series, the resulting group is called an *alkyl radical;* the radical is named by changing the final syllable of the hydrocarbon name from *-ane* to *–yl*.

In order to name paraffin isomers, we must modify our rule. There are, for example, *three* pentanes. In order to have a different name for each of the three structures, the following convention is followed. First, structural formulas are written for the possible arrangements; second, the longest chain is numbered in such a manner that the side radicals are attached to the carbons of the lowest possible numbers; and third, we base the name for each compound on the longest chain.

ACETYLENE SERIES

The name of any homolog of this series is derived from that of the methane hydrocarbon which contains the same number of carbon atoms by replacing the suffix *-ane* by *-yne*. Thus, acetylene, is called *ethyne;* and $CH_3C = CH$, *propyne,* and so on. The triple bond is located in the manner used for the olefins.

AROMATIC HYDROCARBONS

A vast amount of experimental evidence indicates that an astronomical number of organic substances are constituted by molecules which contain a nucleus of 1 or more carbon rings, each ring in turn containing *6 unsaturated carbon atoms.* All hydrocarbons (and their derivatives) of such a constitution are referred to as *aromatic compounds.* The term aromatic has its origin in the fact that certain substances (e.g., oil of bitter almonds, vanilla and oil of wintergreen) are cyclic compounds.

The possession of an odor is not characteristic, however, of all aromatic substances.

It is indeed a difficult task to find an adjective which embraces the significance of the chemistry of aromatic hydrocarbons and their derivatives. It is certainly an exact statement of truth to say that our lives have been revolutionized by discoveries in this area of organic chemistry. "Miracle drugs," modern insecticides (e.g., DDT), dyes, plastics and explosives are but a glimpse of the complete picture. Through the study of complex aromatic structures found in living cells, the biologic chemist is getting ever closer to a reasonable explanation of physical life. The student of the biologic sciences will undoubtedly embellish his knowledge by an understanding of some of the basic principles.

COAL TAR

Benzene, together with some of the other members of the series, is a product of the *destructive distillation* of bituminous coal; i.e., the coal is heated in the absence of air. First, combustible gases escape: then a viscous black liquid distills out, leaving a residue of coke in the retort. This black liquid, called *coal tar,* is probably the most valuable raw material of the chemical industry. When subjected to *fractional distillation,* it yields a conglomeration of aromatic substances; of these, *benzene, toluene, naphthalene, anthracene, phenol and cresol* are the most important. Each of these substances serves as the source material from which thousands of other useful compounds are prepared. All compounds, whether obtained directly from coal tar or indirectly, are commonly referred to as "coal-tar derivatives."

PROPERTIES

The general properties of the series are reflected by those of benzene. Benzene is a colourless liquid boiling at 80° C. It has a peculiar, pleasant odor. Several of the other homologs have a similar odor. Benzene is insoluble in water, but soluble in alcohol and chloroform. It burns with a bright, luminous, smoky flame.The ring structure of the benzene molecule is well established.

POLYMERIZATION

An interesting chemical phenomenon occurs when acetylene gas is passed through a tube heated to dull redness *benzene is formed.* This fact is of fundamental importance: *It shows that a simple substance can unite with itself to produce a more complex molecule; this is called polymerization.* In this case, 3 molecules of acetylene (C_2H_2) unite to form 1 molecule of benzene. Benzene, therefore, may be referred to as a *polymer* of acetylene.

Nature is a master in the art of polymerization. Starch, proteins, glycogen and rubber are but a few of the many important polymers of the biologic world.

Polymers such as these present the most complex substances known to the chemist. Rubber, for example, is a polymer of a substance called *isoprene.*A rubber molecule is formed in the tissues of the rubber tree by the union, or polymerization, of a thousand or more isoprene molecules. The composition of rubber may be represented by the formula in which n stands for a large but indefinite number. This knowledge led to the synthesis of rubber from isoprene and similar substances. Polymerization will be further discussed under carbohydrates and proteins.

OTHER CLASSES OF ORGANIC COMPOUNDS

As stated previously, all organic compounds may be considered as derived from hydrocarbons. These derivatives regardless of their simplicity or complexity may be easily classified on a basis of similar chemical properties. Similar chemical properties, however, result from the presence of a characteristic group, or radical. Actually, therefore, our classification is based on the presence of such a group. This is analogous to our classification of inorganic compounds. It will be recalled, for example, that all acids regardless of the elements from which they are composed liberate the hydrogen ion (H+) in solution; i.e., acids are acids because of the H^+ ion. The following are the principal classes of organic compounds i.e., in addition to the hydrocarbons:

1. Halogen derivatives
2. Alcohols
3. Phenols
4. Aldehydes
5. Ketones
6. Acids
7. Esters
8. Ethers
9. Amines
10. Heterocyclics
11. Metallic derivatives
12. Carbohydrates
13. Fats
14. Proteins

Each of the above classes is characterized by a certain chemical radical. It is only fair to state, however, that there are many organic substances which contain two and sometimes three, but rarely four groups, or radicals, of different classes. In this case, such a compound belongs to more than one class. Often, however, one radical is strong enough to overshadow the properties of the other radical or radicals.

HALOGEN DERIVATIVES

A large number of compounds may be formed from hydrocarbons by the substitution of halogen elements (i.e., fluorine, chlorine, bromine and iodine) for hydrogen atoms. In the case of unsaturated hydrocarbons, *addition* takes place. Although it is possible to form such compounds through direct action of the halogen on the hydrocarbon, it is usually a much simpler matter to do this by other methods.

The following halogen derivatives will serve to illustrate this class and, in addition, indicate the value of certain members.

CHLOROFORM ($CHC1_3$)

Chloroform is a heavy, colourless, sweet smelling and noninflammable liquid possessing potent anesthetic properties. Although beset by toxic actions on heart and liver tissues, it still enjoys some use in this area. Chloroform is an excellent solvent for fats, oils and resins.

CARBON TETRACHLORIDE

Carbon tetrachloride is one of the most important industrial and commercial organic compounds. It is a noninflammable, colourless liquid with a low boiling point. Its chief property is its ability to dissolve grease and fat readily; thus it is a safe liquid for dry cleaning. It also is the only liquid employed to extinguish electrical fires. When sprayed on a fire, it vaporizes to form a heavy gas which excludes oxygen.

lodomethanes. The most important iodine derivative of methane is triiodomethane, commonly known as *fodoform* lodoform is a yellow powder with an extremely characteristic odor (perhaps best described as "medicinal").

lodoform still enjoys some use as an antiseptic dusting powder. Todoform is readily prepared by the action of iodine and sodium hydroxide on ethyl alcohol. This reaction is often used as a *test* for ethyl alcohol.

HALOGENATED AROMATICS

Just about any halogen derivative desired can be prepared from the aromatic hydrocarbons. The reaction of benzene with chlorine will serve as an example. Of the chloro-benzene derivatives, p-dichlorobenzene is of commercial importance. From our previous work with isomerism, we know that there are two other chlorobenzenes *ortho* (o-) and *meta* (m-).Only the para form, however, possesses insecticidal properties.

TEST FOR SATURATION

Bromine and iodine are frequently employed in the laboratory to determine if a compound is saturated or un-satur ited. It will be recalled that unsaturated compounds are much more reactive than saturated compounds. When, for example, a solution of bromine is added to an unsaturated compound, the reddish-brown colour of the solution quickly disappears; i.e., an addition product is formed. Saturated substances react at a much slower rate and sometimes not at all.

IDENTIFICATION

Halogenated organic compounds may be identified in a rather crude but fairly accurate fashion in the following manner. A copper wire is dipped into the compound, and then held in the flame of a Bunsen burner. A positive test is indicated by a *green* flame.

GENERIC FORMULA

Throughout our work with the various classes of organic compounds, we shall find it convenient to employ a general, or a so-called *generic formula*. The generic formula for halogenated derivatives may be given by *RX*, where R stands for an *alkyl* or *aryl* group, and X for the halogen element.

ALCOHOLS

Alcohols comprise a large and important class of organic compounds. An alcohol is a substitution product of a hydrocarbon in which a hydrogen atom of the hydrocarbon has been replaced by a *hydroxyl group* (OH). An alcohol may, therefore, be considered as an *organic hydroxide*, or *base*.

There are three classes of alcohols: *primary, secondary* and *tertiary*. A primary alcohol is an alcohol in which the hydroxyl group is attached to a carbon atom which, in turn, is attached to not more than 7 other carbon atom; a secondary alcohol is one in which the hydroxyl group is attached to a carbon atom which, in turn, is connected to 2 carbon atoms; and a tertiary alcohol is one in which the carbon atom is attached to 3 other carbon atoms.

TERTIARY ALCOHOL

Although all alcohols share common properties, the behaviour of any one alcohol depends to some degree upon whether it is primary, secondary or tertiary.

Alcohols are known which contain two, three, four, five or six hy-droxyl groups; such alcohols are referred to as polyhydroxy alcohols.

The following alcohols are the most important of the entire class. Moreover, they will serve to illustrate the general behaviour of this large group of compounds.

METHYL ALCOHOL

Methyl alcohol (CH_3OH), also known as *methanol* or *wood alcohol,* is the simplest alcohol possible. It is one of the volatile products formed when wood is subjected to destructive distillation. Under proper conditions carbon monoxide and hydrogen unite and form methyl alcohol.

Methyl alcohol is a colourless liquid which boils at 65° C. It mixes with water in all proportions and is an excellent solvent. Many substances which are insoluble in other liquids are soluble in methyl alcohol. The alcohol burns with a pale blue flame which does not deposit soot.

Chemically, methyl alcohol reacts with several substances. It reacts with acids, for example, to form an important group of compounds called *esters.* We shall discuss esters in a subsequent section.

Methyl alcohol is used extensively commercially as a solvent for shellac, in making varnish, and in the preparation of dyes and formaldehyde. In addition, it is a valuable solvent and reagent in the laboratory. Methyl alcohol is also used in biology as a preservative for specimens. OL vital toxicologic interest is the fact that methyl alcohol is a potent poison, causing partial or complete blindness in small doses. Its inherent toxicity coupled with its availability makes methyl alcohol a dangerous article in the laboratory, industry and household. There are numerous poisonings among those who ingest methyl alcohol as a substitute for "alcohol" (i.e., grain alcohol). There are many such persons, however, who are not aware of the difference. Each year there are about 150 deaths resulting from this substance.

Methyl alcohol is not as depressant to the central nervous system as grain alcohol. It differs from ordinary grain alcohol also in that it is not completely oxidized in the body to carbon dioxide and water. As a result, intermediate products *formaldehyde* and *formic acid* are produced. It is thought that these products are the cause of the blindness.

ETHYL ALCOHOL

Ethyl alcohol also known as *ethanol, grain alcohol* or *"alcohol"* is unquestionably the most infamous and notorious compound known to man. There seems to be no other substance which has enlisted as much comment from the sages of chemistry, biology and philosophy past and present. Alcohol at least its properties has been known since the earliest period of written history perhaps before. The ancient Egyptians, for example, were well aware of its spiritous and preservative properties.

Ethyl alcohol is produced by the oldest chemical process known to man *fermentation*. Fermentation is defined as the degradation of carbohydrates (i.e., sugars and starches) by organic catalysts called *enzymes.*

Alcoholic fermentation is carried out by adding yeast to a warm solution of glucose ($C_6H_{12}O_6$). The yeast contains an enzyme called *zymase* (produced by yeast cells) which brings about the following reaction:

$$C_6H_{12}0_6 = 2C_2H_5OH + 2CO_2, \text{ glucose alcohol}$$

Cane sugar is also fermented by yeast. Since zymase can only act upon simple sugar, i.e., glucose, the yeast must contain some other enzyme which converts cane sugar (sucrose) to glucose. This enzyme is called *invertase*

Starch, however, is the cheapest source material for producing ethyl alcohol and spiritous beverages. In this process malt (which is made from sprouting barley) is added to a mixture of some grain (source of starch) and water. An enzyme in the malt called *diastase* converts starch into *maltose,* an isomer of sucrose. Yeast is then added. The yeast acts upon the maltose in the manner described for sucrose. Alcohol is obtained from the mixture by fractional distillation. By distillation it is possible to obtain 95 per cent alcohol. When 100 per cenFalcohol *(absolute alcohol)* is desired, it is necessary to treat the 95 per cent with a dehydrating agent such as lime.^ In recent years, a totally synthetic process of producing ethyl alcohol has been adopted in some areas. The synthetic process starts with ethylene and sulfuric acid.

Pure ethyl alcohol has a peculiar, pleasant odor. It boils at 78° C. It burns with a nonluminous flame which does not leave a deposit of soot. Its effects upon the body are a result of its depressant action upon the central nervous system. When taken in large doses it may cause death. Chemically, ethyl alcohol reacts with a number of different chemical reagents. The products formed contain the *ethyl radical* (C_2H_5). Later on we shall mention a few important ethyl esters.

Millions of gallons of alcohol are used annually in various types of industry. As a solvent, ethyl alcohol stands next to water. There are numerous pharmaceutical preparations, e.g., tinctures, spirits and elixirs, which contain medicaments dissolved in alcohol. Of course, all spiritous beverages contain ethyl alcohol as the "active" constituent. In order to provide alcohol for industrial use at low cost (i.e., *tax-free),* the government permits the sale of denatured alcohol. Alcohol is denatured by adding to it substances, such as *methyl alcohol* or *pyridine,* which render the alcohol unfit for human consumption.

GLYCERIN

Glycerin is a heavy, colourless liquid, soluble in water and sweet-tasting. A sweet taste is characteristic of polyhydroxy alcohols. Glycerin is a by-product of the soap industry. Commercially, glycerin enjoys wide use. It is employed in the manufacture of nitroglycerin, and in the preparation of plastics, toilet articles and pharmaceutical preparations. Since glycerin is a constituent of practically all animal and vegetable fats and oils, it is of biologic importance. The chemistry of glycerin will be discussed in other sections.

NOMENCLATURE

Simple alcohols are named in the manner indicated above (e.g., *methyl alcohol, ethyl alcohol,* and so on). A more systematic and accurate way is as follows: The name of the alcohol is formed by replacing the final *e* in the name of the hydrocarbon from which the alcohol is derived by the ending *ol* (e.g., methyl alcohol is *methanol,* and ethyl alcohol is *ethanol).* When necessary, the position of the hydroxyl group is shown by adding to the name of the alcohol a number which indicates the position of the group in relation to the carbon atoms in the chain.

Alcohols which contain two or more hydroxyl groups are named in a similar way. In this case, however, the number of hydroxyl groups is indicated by prefixing the syllables *di-, tri-*, and so on, to the suffix *-ol.* Alcohols with one hydroxyl group at the end, i.e., at the number 1 position, are often referred to as *normal.*

PHENOLS

The hydroxyl derivatives of the aromatic hydrocarbons are of two distinct classes: those in which the hydroxyl group is situated in a side chain; and those in which the hydroxyl group is linked *directly* to the ring. The -former class closely resembles the aliphatic alcohols, while members of the latter class, which are called *phenols,* possess acid properties. Benzyl alcohol, for example, is a "regular" alcohol because the hydroxyl group is linked to the *side chain.*

Actually, this alcohol may be considered as a substitution product of methyl alcohol, formed is the result of the replacement of 1 hydrogen atopi by a phenyl radical. This section, then, is only concerned with phenols. The chemistry of this class may be well illustrated by a consideration of phenol and cresol.

Phenol commonly known as *carbolic acid,* occurs in coal tar. It is a colourless substance which crystallizes in long needles which melt at 41° C. Phenol possesses a strong characteristic odor, described as "phenolic." Phenol dissolves in about 15 parts of water.

Phenol is quite reactive, and forms a number of important derivatives. Its acid properties are disclosed by the fact that it readily reacts with bases to form salts. Phenol is one of the most important source materials obtained from coal tar. Several of its derivatives, e.g., *picric acid* and *salicylic acid,* are of commercial value.

It reacts with formaldehyde to form *Bakelite* and similar plastics. Although inferior to other antiseptics available at the present time, it continues to enjoy some use as a disinfectant and germicide. Of historical importance is the fact that phenol was the antiseptic used by Lister to herald "antiseptic surgery." Phenol is still the standard by which other antiseptics, disinfectants and germicides are evaluated; i.e., for every such agent a *phenol coefficient* is ascertained. A disinfectant with a phenol coefficient of 2, for example, means that it is twice as effective as phenol.

Phenol is of considerable toxicologic concern. Although less popular than in former years, it still remains high on the list of suicidal agents. Its toxic actions are those of a powerful corrosive.

CRESOL

Whereas phenol is derived from benzene, cresol is derived from toluene. There are three isomeric cresols. They are all found along with other phenols in *creosote.* (Creosote is a mixture of chemical compounds contained in coal tar and the tars from pine and beechwood.) Although the isomers may be purchased separately, they appear on the market as a mixture. Unless otherwise specified, the term "cresol" refers to the mixture.

The properties of cresol resemble closely those of phenol. Cresol finds its most important use as a disinfectant and germicide. In this capacity it is superior to phenol. *Lysol* and similar products are essentially soap solutions of cresol. The soap is added to enhance the solubility of the cresol in water.

OTHER PHENOLS

Other important phenols include: *thymol, naph-thols, catechol resorcinol, hydroquinone* and *pyrogallol.*

IDENTIFICATION

The reactions of the phenols and their derivatives with *alkalis, ferric chloride* and *bromine water* are of particular value in their identification. Phenols react with alkalis to form soluble salts. Ferric chloride produces marked colourations in aqueous solution. With bromine water most phenols yield a precipitate of a brominated derivative.

ALDEHYDES

Aldehydes are derived from *primary alcohols by partial oxidation.* This fact is of great theoretical and practical importance. Thus, from a given alcohol, we can prepare the corresponding aldehyde. The most common and characteristic members of the class are *formaldehyde* and *acetaldehyde.*

FORMALDEHYDE

Formaldehyde (H-CHO) is obtained through the oxidation of methyl alcohol. When methyl alcohol vapour and air are passed over a hot copper gauze, formaldehyde gas and water are formed.

Formaldehyde is the simplest aldehyde. It is a gas which is soluble in water, and has a powerful, irritating odor. The chemistry of formaldehyde is interesting. Formaldehyde is readily oxidized to *formic acid* and carbon dioxide. It is readily reduced by nascent hydrogen to methyl alcohol.

With ammonia, formaldehyde reacts to form a complex substance called *hexamethyleneielramin.*This compound is used in medicine under the name "methenamine" as a urinary antiseptic.

In the presence of a suitable catalyst, formaldehyde readily *polymerizes* to form a white solid called *metaformaldehyde*. There is some evidence that the formation of starch in green plants (photosynthesis) involves a similar reaction. It has been suggested that formaldehyde is first formed from carbon dioxide and water (in the presence of chlorophyll and sunlight), and that the aldehyde then undergoes polymerization to produce carbohydrates.

Formaldehyde comes on the market as "formalin," which is a 37 per cent aqueous solution of formaldehyde gas. The words "formaldehyde" and "formalin" are often used synonymously. At the present time formaldehyde seems to be the more popular term. Huge quantities of formaldehyde are used in the manufacture of certain dyes, plastics (e.g., *Bakelite, Formica)* and photographic films. In the biologic laboratory it is extensively used as a preservative for anatomic specimens. In the hospital it still enjoys some use as a disinfectant. Formaldehyde was formerly employed as a food preservative. This practice is now prohibited by law.

ACETALDEHYDE

Acetaldehyde is prepared by the oxidation of ethyl alcohol. The oxidation is readily carried out by a mixture of sulfuric acid and potassium dichromate. Acetaldehyde, commonly called *acetic aldehyde,* or simply *aldehyde,* is a colourless liquid, boiling at 21° C. It mixes with water and alcohol in all proportions. Its odor is marked and characteristic. From the chemical point of view, the most characteristic property of aldehyde is its power to unite directly with other substances.

Indeed, if left to itself, it readily changes into polymeric modifications; of these *paraldehyde* is the most important. The paraldehyde molecule is formed by the union of 3 aldehyde molecules; its formula is $(CH_3CHO)_3$. Paraldehyde has been used for years as a *hypnotic.* Another interesting compound of acetaldehyde is *chloral.* Chloral is formed by the substitution of the 3 hydrogens of the methyl group with 3 chlorine atoms.

Chloral is a colourless liquid which boils at 98° C. When chloral and water are brought together, they unite to form a crystallized compound *chloral hydrate.* In doses of 0.5 gm. to 2 gm., chloral hydrate produces a natural sleep; in larger doses it acts as an anesthetic.

Nomenclature. The simple aldehydes are usually named from the acids which they yield on oxidation; thus H'CHO is *formic aldehyde* or *formaldehyde,* and CH.CHO is *acetic aldehyde* or *acetaldehyde.* In the more systematic method, the name of an aldehyde is formed by adding the suffix *-al* to the name of the hydrocarbon which contains the same number of carbon atoms. According to this system acetaldehyde is *ethanal.*

Identification. The ability of aldehydes to reduce other substances (i.e., they arc reducing agents) is of analytical significance. When an aldehyde or compound containing an aldehyde group is added to an *alkaline solution of a cupric salt, a red precipitate of cuprous oxide* (CuO) *is formed.*

Chapter 9

Biochemistry

Biochemistry is a marriage of the various branches of chemistry especially organic and biology; it may be defined, therefore, as the *chemistry of life.* In this chapter we shall direct our attention to the chemistry of the various physiologic systems of the human body. In general, of course, our study will apply to all plant and animal tissues.

PROTOPLASM

The difference between something that is living and something that is dead will probably always appear as a mystery, even to the most brilliant of minds. This is not to say that the scientist will never be able to explain what life is, but rather that he will not be able to comprehend what he has explained. For example, the astronomer can explain the nature of the universe to an unbelievable degree of detail, yet he views the myriad of stars above his head with the same wonderment as the savage.

As a matter of fact, the biochemist and the astronomer have a great deal in common: They are both dealing with something infinitely distant from their powers of visualization. Scientifically, therefore, it seems prudent to think of life i.e., protoplasm as a concentrated universe of physical and chemical forces operating under the direction of energies derived from their unique synthesis. This statement represents the best attempt of the author to convey, in a *practical* way, the thoughts of the biochemist.

Now let us consider protoplasm on a pure physical and chemical basis. Protoplasm is the "stuff" of life. Specifically, it is the term used to describe the material which constitutes the living cell plant or animal. The composition of protoplasm is not a mystery; as a matter of fact, we know a great deal about its constituents. Protoplasm is a *viscous, colloidal sol which contains proteins, carbohydrates, lipids, salts and enzymes dispersed or dissolved in water the chief constituent.*

That protoplasm is not a simple mixture of these substances is demonstrated by the fact that it carries on *metabolism;* i.e., it consumes oxygen and liberates carbon dioxide and energy. If metabolism stops, our mixture is no longer protoplasm because it is "dead."

The composition of the "living mixture" and the "dead mixture," however, is the Biochemistry is a marriage of the various branches of chemistry especially organic and biology; it may be defined, therefore, as the *chemistry of life.* In this chapter we shall direct our attention to the chemistry of the various physiologic systems of the human body. In general, of course, our study will apply to all plant and animal tissues.

ENZYMES

The complex chemical reactions within (and without) the cells of the body are made possible by a class of catalytic substances called *enzymes.* Thus, an enzyme may be defined as an *organic catalyst.* It is now known that all forms of life depend upon these agents for their existence. Yeast cells, for example, derive their energy through the agency of the enzyme, *zymase.* Zymase converts glucose into alcohol, carbon dioxide and energy.

A number of enzymes have been isolated in the pure crystalline state. Extensive biochemical research has shown that they are *proteins* or proteins joined on to other groups.

Enzymes are also characterized by their extreme *potency* and *sped-ficity.* A few grains of trypsin, for example, will easily convert a pound or two of meat (protein) into amino acids, whereas this enzyme has no action on carbohydrates. Enzymes are very sensitive to their environment, especially to pH and temperature changes. The enzyme ptyalin, for example, will not act in a basic medium; trypsin, on the other hand, requires a basic medium. As for temperature, they function best at body heat.

At the present time enzymes are named by adding the suffix *-ase* to a stem derived from the substance (called a "substrate") acted upon. For example, *maltase* converts *maltose* to glucose. Old names, however, such as pepsin, ptyalin and trypsin, are still used. Enzymes may be classified on a basis of the *types* of substrates. For example, the term *carbohydrase* could be applied to all enzymes which act upon carbohydrates. Thus, maltase would be a carbohydrase. The following classes of enzymes, together with representative members, are of physiologic importance. At appropriate places, the individual members will be discussed.

Esterases

Esterases are enzymes which hydrolyze organic esters. Examples: *pancreatic lipase* and *phosphatases.*

Proteinases

Proteinases hydrolyze proteins to amino acids and intermediate products. Examples: *pepsin, trypsin, rennin* and *papain.*

Peptidases

Peptidases hydrolyze polypeptides to amino acids. Examples: *aminopolypeptidase, carboxypolypeptidase* and *dipeptidase* (erepsin).

Phosphorylases

Phosphorylases bring about the hydrolysis and synthesis of polysaccharides.

Carbohydrases

Carbohydrases hydrolyze carbohydrates to simpler products: Polysaccharides and disaccharides are converted to mono-saccharides. Examples: *sucrase, maltase, lactase, ptyalin, pancreatic amylase.*

Oxidases

Oxidases play a vital role in so-called biologic oxidations, i.e., the utilization of oxygen within the cell. This subject will be discussed later. Important oxidases include: *catalase, dehydrogenase, per-oxidase* and *cytochrome oxidase.*

ESSENTIALS OF LIFE

In order for protoplasm to survive, grow and reproduce itself, it must be supplied with the following chemical substances, or *nutrients: carbohydrates, fats, proteins, minerals, vitamins, water and oxygen.*

WATER

Water is the basic constituent of protoplasm; all the other constituents are either *dispersed* (in a colloidal fashion) or *dissolved* in it. The bulk of plant and animal tissues, therefore, is due to water. Such vegetables as tomatoes and cucumbers are almost 100 per cent water; animal tissues contain in the vicinity of 70 per cent water.

The water of the body is said to be present in two compartments: *intracellular* and *extracellular*. Intracellular water refers to the water within the cell; extracellular water refers to the water outside the cells specifically, in the *blood, lymph and tissue fluid.* It is now generally accepted and appreciated that the balance the so-called *water balance* between these two compartments, and within the extracellular compartment, is one of the most critical situations in physiology.

This is so because just about every physical and chemical action within our tissues indirectly affects the *equilibrium* (balance) between the locations mentioned above. For example, *edema* (the accumulation of water in the tissues) and *dehydration* (the loss of intracellular water) attend a large number of pathologic states. In general, any condition which upsets the *composition* (especially of *metallic ions* and *protein)* or the *pressure (osmotic, hydrostatic,* etc.,) of blood will bring about a shift of water.

MINERALS

In nutrition the term *minerals* refers to the salts (usually inorganic) of essential *metals* and *nonmetals,* viz., K, Na, Ca, Mg, Fe, Cu, S, P, F, Cl, I, SO,, HCO,. A well-balanced diet supplies an ample amount of such substances. Metallic and nonmetallic ions are present within the cells (i.e., dissolved in intracellular water), and in all the fluids of the extra cellular compartment. Although present in relatively small amounts, they play important roles in the distribution of the body's water,-the action of enzymes and the formation of bones and teeth.

The cells of the body are particularly sensitive to the *ratio* between the ionic concentrations (viz., Na+, K+, Ca^{+}+). Any single ion is *toxic* to cells, but this toxicity is antagonized by the presence of others *ion antagonism.* In order for the body's tissues to function properly, therefore, the *ionic balance* (commonly referred to as *electrolytic balance)* must be maintained.

Extensive research has indicated that a great many pathologic manifestations are directly due to a shift in the ionic concentrations. Already the principles of electrolyte balance are being employed in medicine. It is now realized that practically all patients especially those at "low ebb" are benefited by the intravenous adminis tration of electrolyte solutions (NaCl, KC1, $CaCl_2$) that have been

carefully calculated to meet the ionic needs of the patient's body.

OXYGEN

Oxygen is needed for the oxidation of fuels within the cell to furnish energy for life's functions. Our present information on this subject definitely shows that this is not a conventional type of "burning." As a matter of fact, the way in which energy is gained by the cell has proven to be one of nature's most uncanny and cunning chemical tricks. The important details will be given under the topic of respiration.

VITAMINS

Since the beginning of the century it has been realized that certain nutritional factors other than water, minerals, carbohy drates, fats and proteins are essential to animal and human life. Concrete ideas, however, were not available until Hopkins in 1906 demonstrated that normal foods do contain *accessory factors*. In 1911 Funk, who believed that these accessory factors were organic *amines* (i.e., containing the NH_2 group), suggested the term *vitamine*. Further work by many investigators disclosed that the amino group was not character istic. In 1920 Drummond proposed a compromise, using *vitamin* without the final *e*.

The role that vitamins play in the welfare of man represents one of the most significant contributions of the chemist. It is now known with mathematical precision how much of most vitamins we must ingest in order to be healthy and free from the several *deficiency diseases*. The mechanisms by which the various vitamins perform their miracles were long wrapped in a shroud of mystery. Today the biochemist is able to offer an almost complete explanation. In general it may be said with some degree of accuracy that vitamins are the *regulators* of the chemical reactions within the cell. Specifically, they are concerned with the action of the cell's enzymes. Most vitamins of the B complex, for example, are known to function as enzyme helpers, or *coenzymes*.

All the vitamins have been isolated in the pure state. This has led to the determination of their chemical structure, and finally to their synthesis in the laboratory. Physically, the vitamins are often categorized according to their solubility in water or fat. The so-called *water-soluble vitamins* include vitamin C and members of the B complex group; the rest A, D, K and E are *fat-soluble*.

FATS, CARBOHYDRATES AND PROTEINS

In order to be of use, fats, carbohydrates and proteins must be digested. The chemical aspects of this process are described under *digestion*. The digestive derivatives of carbohydrates are used by the cell to furnish *energy;* fats supply energy and form *adipose* (fatty) *tissue;* proteins are employed to *synthesize new protoplasm,* i.e., to repair and build tissue.

DIGESTION

Vitamins, minerals, water and oxygen are able to enter a cell directly; fats, carbohydrates (except monosaccharides) and proteins are not able to do so. In order for these substances to be absorbed and assimilated (i.e., synthesized into protoplasm), they must be *digested,* i.e., split into smaller soluble molecules. Thus, proteins are hydrolyzed to *amino acids;* carbohydrates are hydrolyzed to *glucose, fructose* and *galactose;* and fats are hydrolyzed to *fatty acids* and *glycerin.* The digestion of these substances is carried out in the mouth, stomach and small intestine through the agency of digestive *enzymes* secreted by a number of glands along the way. The details of the digestive process are presented below.

The Mouth. The chief function of the mouth in the digestive process is mastication (chewing). Mastication reduces the size of the food particles for ease in swallowing, and facilitates enzyme attack throughout the alimentary canal. Mastication, of course, would be impossible without the moistening and lubricating effects of saliva. Saliva is secreted into the mouth by the sublingual, submaxillary and parotid glands (the so-called *salivary glands).* It contains mucin, salts, water and the enzyme *ptyalin.* Saliva has a pH between 6 and 7. The only chemical action in the mouth is the hydrolysis of starch by ptyalin.

Ptyalin converts starch first into *dextrins,* and then into *maltose.* This action is slight, however, because most of us swallow our food long before the time required for complete digestion. Fortunately, starch is completely digested in the small intestine.

The Stomach. The stomach acts as a storage place for food, and as a partial digester of proteins. These functions, however, are not indispensable; following total gastrectomy a patient seems to be able to take care of his food fairly well, aside from the fact he usually is not able to eat as much as before. Chemical digestion is performed by the so-called *gastric juice.*

This juice consists of *water, salts, enzymes, mucin and hydrochloric acid.* Due to the presence of hydrochloric acid, it is highly acid. During fasting, the *pH* is about 3.5; following stimulation of secretion by food, the juice has a *pH* of about 1. Of interest is the fact that this juice is secreted by three types of cells. The *parietal cells* (located in the fundus and body of the stomach) secrete the HC1; the *mucous cells* (located throughout the stomach) secrete mucin; and the *zymogenic cells* (also found throughout) secrete the enzymes.

Most texts state that there are three enzymes present in the juice: *pepsin, rennin* and *gastric lipase.* It is quite safe to say that gastric lipase is unimportant. Rennin is equally unimportant in the highly acid stomach of the adult. In the less acid stomach of the infant, however, rennin is important in clotting milk.

The chief chemical action of the juice, then, is due to the action of pepsin. It is now known that pepsin is secreted in an inactive form called *pepsinogen,* and that pepsinogen is converted to *pepsin* in the presence of HC1. Moreover, pepsin itself requires an acid medium. This can easily be demonstrated in the laboratory. Pepsin is a *proteinase;* it converts proteins to *metaproteins,* soluble *proteoses* and *peptones* intermediate products of protein hydrolysis.

As food leaves the stomach in the form of a semiliquid called *chyme* the proteins are at various stages of digestion. It is highly doubtful that all the proteins of an average meal are converted to peptones; indeed, some may leave unaltered.

A word should be said about *mucin.* Mucin the characteristic constituent of mucus (found throughout the alimentary canal) performs an important function in the stomach: It protects the gastric and duodenal mucosae from mechanical erosion and neutralizes acid. (For these reasons gastric mucin has been used in the treatment of peptic ulcer.)

THE SMALL INTESTINE

It should be quite clear by now that salivary and gastric digestion contribute little in the way of chemical action. Thus, it is possible for carbohydrates, fats and proteins to enter the duodenum in an unaltered condition. The mechanical action of the mouth, stomach, saliva and gastric juice, of course, is unquestionably important; together they reduce the meal to a consistency (chyme) easily attacked by intestinal enzymes. The small intestine has two functions: the *digestion* of carbohydrates, fats and proteins; and the *absorption* of the products of digestion, vitamins, mineral salts and water. Absorption is made possible by its extensive length (20 ft.), and by its unique architecture; both of these factors afford a tremendous suriace area. The digestive juice in the intestine is formed from the secretions of the *liver* (bile), *pancreas* and the *intestine* itself.

PROTEINS

Proteins and the intermediate products of gastric digestion proteoses and peptones are acted upon by *trypsin* to yield peptides and some amino acids. Trypsin results from the action of *entero-kinase* on *trypsinogen* (inactive trypsin). The peptides are then hydro-lyzed to amino acids by *erepsin*. (Erepsin, at one time considered a single enzyme, is now known to be an impure enzyme which contains peptidases.)

The final stage in the digestion, or hydrolysis, of a protein is the splitting of a dipeptide, i.e., two amino acids held together by a peptide bridge. *1 molecule of fat yields 3 molecules of a fatty acid and 1 molecule of glycerin.* Bile is a valuable adjunct in the digestion of fats. The function of bile, by means of bile salts, is to *emulsify* fats in water, thus increasing the surface area for attack by lipase.

LARGE INTESTINE

The chief functions of the large intestine are: *storage* of feces until defecation, and *water absorption* to prevent loss. Water absorption reduces the feces (indigestible food, bacteria, desquamated cells, etc.) to their typical consistency.

It is normal for bacteria to be present in considerable numbers in the colon, where conditions are favourable for their growth. Bacteria play a useful role in the *synthesis of certain vitamins* (e.g., vitamin K) and in the decomposition of substances (e.g., cellulose) which have escaped attack by the digestive juices. Unfortunately, however, certain types of bacteria secrete enzymes that can decompose monosaccharides and amino acids into harmful substances. Carbohydrates are broken down into *acids* (viz., *butyric, lactic, acetic)* and *gases (hydrogen, methane* and *carbon dioxide).*

The action of bacteria on amino acids yields such substances .as *histamine, phenol, indole* and *skatole.* These substances are very toxic if absorbed in considerable amounts. Fortunately, the body is protected against this type of insult by the detoxifying action of the liver. For example, indole is changed to the nontoxic *indican,* which is excreted by the kidneys. The formation of feces completes the digestive process.

ABSORPTION

The absorption of the soluble end products of digestion occurs throughout the mucosa of the small intestine. Ammo acids, glucose, fructose and galactose are absorbed into the capillaries of the villi. Concerning the absorption of fats, there has always been a little difference of opinion. Recent information suggests that *bile salts* form a *soluble complex* with the insoluble fatty acids, thereby permitting them to penetrate the villi. Inside the villi the complex releases its fatty acid, which then unites with glycerin (being soluble glycerin passes into the villi unaided) to form fat. Once fat has been resynthesized inside the villus it is taken into the lacteals, and eventually enters the blood stream via the lymph. The mechanism of absorption involves more than the physical forces of diffusion and osmosis; it is probably an *active secretion* by the cells of the mucosa. This is supported by experimental evidence that certain closely related substances are absorbed at *different* rates. For example, glucose is absorbed much more rapidly than fructose.

Once the nutrients are absorbed into the blood they are transported to the body's tissues, where they diffuse through the thin capillary walls into the fluid surrounding the cells *(tissue fluid).* The nutrients then pass through the cell's *selective* membrane. Here again, it seems to be more than a simple case of diffusion or osmosis; the cell membrane seems "to know" what it wants to let in and out. For example, how else can we explain the fact that certain optical isomers are *physiologically different?* The cell membrane, then, functions on a basis of its ability to accommodate some structures arid exclude others. In addition, recent work in this area indicates that the membrane expends energy in the transport process.

INTRAVENOUS ALIMENTATION

Food by vein is now a standard procedure in medical practice. Solutions are available which supply all the essential nutrients in a balanced form. Carbohydrate is administered as *glucose, fructose* or *invert sugar;* protein is given as *amino acids* (prepared by hydrolyzing "whole proteins," i.e., those that contain all the essential amino acids, such as casein, with digestive enzymes); electrolytes (viz., NaCl, KC1 and $CaCl_2$) and vitamins are given in simple solution. These various solutions are given singly or combined.

Metabolism. *Metabolism may be defined as all the chemical changes which take place within the tissues.* Metabolism may be subdivided into *anabolism* and *catabolism.* Anabolism refers to all chemical changes which participate in the *synthesis* of protoplasm *(assimilation),* or other vital substances; catabolism refers to all chemical changes which involve the *degradation* of cellular constituents, yielding *waste products* and *energy.* In this section we are interested in the metabolism of the digestive products. The main points are briefly stated below.

CARBOHYDRATES

As we already know, carbohydrates make their way into the body in the form of the monosaccharides *glucose, fructose* and *galactose.* When the blood passes through the liver these three sugars when present in excess amounts are stored as the polysac-charide *glycogen (glycogenesis),* which can be reconverted into glucose *(glycogenolysis)* and given up to the blood as needed. In this way, the glucose concentration of the blood remains constant.

The chief function of glucose in the tissues is to supply *energy*. This involves the oxidation of glucose to carbon dioxide and water. The reaction may be stated as follows:

$$C_6H_{12}0_6 + 60_2 = 6C0_2 + 6H_20 + energy$$

Actually, the above reaction is only a summary of an extremely complex process. A rather complete analysis of the intermediate steps will be given under *cellular respiration.*

Unfortunately for those of us who have a tendency to put on weight, *excess glucose can be converted into fat.* Although this fact has long been recognized, the chemistry of the transformation has only recently been elucidated. It is believed that certain of the metabolic products derived from glucose (during its oxidation) are converted into *fatty acids,* and that these fatty acids combine with *glycerin formed by the reduction of glucose* to yield fats, which are stored in the fatty deposits of the body.

FATS

The liver is essential to the metabolism of fats. The fats which are absorbed into the blood from our food must be changed into a type which is able to enter the cell. A good portion of fat is transformed into *phospholipids* (via the liver), which the cells seem to absorb selectively from the blood. Fats (fatty acids) are also present in the blood as esters of *cholesterol.*

Once fats have been taken into cells, they may be oxidized to yield energy, or stored as fatty tissues. The oxidation of fats, like the oxidation of glucose, is not a simple process. The essential details will be presented when we consider cellular respiration.

PROTEINS

Proteins are taken into the blood as amino acids. Following absorption by the tissues they serve chiefly as the building blocks of *protoplasm, enzymes, hormones* and *blood proteins.* In excess amounts, however, amino acids undergo catabolic reactions, yielding *glucose* (and indirectly fat) *energy* and *nitrogenous waste products, viz., urea.*

The first of these catabolic reactions *oxidative deamination* is carried out in the liver. Deamination means the splitting off of amino groups.

BLOOD (CIRCULATORY SYSTEM)

The unity of the organism is maintained by the circulation of a liquid tissue referred to as blood. There are other vital liquids in the body, however, in addition to blood; these include *tissue fluid, lymph, cerebro-spinal fluid, synovial fluid, aqueous humour* and *fluids of the ear* (endo-lymph and perilymph). It is important to realize, though, that all of these fluids are directly or indirectly derived from blood. For example, tissue fluid is a direct filtrate of blood, while lymph is derived from tissue fluid. In this section, we shall discuss the chemistry of blood, tissue fluid and lymph.

Blood is a viscous fluid composed of 45 per cent cells and 55 per cent plasma.The cellular elements include *erythrocytes, leukocytes* and *platelets,* or *thrombocytes.* The liquid portion, or plasma, is a complex aqueous solution of *proteins, hormones, vitamins, lipids, gases, waste products* (viz., *urea)* and *inorganic salts.*

Blood performs a variety of vital tasks: It supplies the cells with the essentials of life; removes waste products; carries hormones; regulates the concentration of ions; regulates the osmotic pressure, temperature and gas tensions of tissue fluid; forms tissue fluid; aids in the defence against infection via leukocytes and antibodies. Of concern to us, how ever, are the important biochemical aspects of blood. The following topics are perhaps those of chief interest to the student of physiology.

Hydrogen Ion Concentration

Human protoplasm is killed unless its hydrogen ion concentration, or pH, remains between 7.2 and 6.7; nervous tissue is especially sensitive to a change in pH. There are two or three tissues, however, which can withstand a higher or lower pH. For example, the cells of the urinary tract can be subjected to urine of pH 4.8 to 8; and the gastric and duodenal mucosae are bathed with HC1 of pH 1, although these are undoubtedly protected by a coating of mucus. A shift in the pH probably produces its deleterious effects by creating an unfavourable environment for cellular enzymes, thus interfering with the cell's chemistry; and by denaturing proteins, i.e., causing them to swell. The pH of protoplasm is maintained by the *tissue fluid* which, in turn, is maintained by the *blood* through various independent biochemical and physiologic mechanisms: *blood buffers, respiration, kidneys* and *alimentary tract.* Blood itself has a *pH* of 7.43 to 7.4 (arterial), and 7.4 to 7.35 (venous).

BUFFERS

A buffered solution is a solution which within reasonable limits resists a change in pH. A buffer neutralizes strong acids and bases by the formation of *weaker* ones. The blood buffers include *bi-carbonates, phosphates* (especially their Na salts) and *proteins* (viz., albumin, globulin and hemoglobin).

There are several pathologic conditions in which an abnormal amount of acid substances escape into the blood. Usually the buffers (especially bicarbonate) are able to neutralize this excess acidity at a *constant pH.* Sometimes, however, the blood's buffer system is taxed beyond its powers, and death results if the pH goes below 6.9. An increase in the acidity of the blood is called *acidemia,* or *acidosis.* If the buffer system is able to cope with the situation without a change in pH, the term *compensated acidemia* is used; if there is a change in pH, it is called *uncompensated acidemia.*

A decrease in the hydrogen ion concentration of the blood (i.e., an increase in pH) is called *alkalemia, or alkalosis.* Like acidemia, it may be *compensated* or *uncompensated.* Since the metabolic waste products of protoplasm are usually acidic, alkalemia is not as common as acidemia. There are several conditions, however, which will provoke alkalosis. The chronic ingestion of $NaHCO_3$, for example, can bring about this condition.

RESPIRATION

Respiration controls the pH of the blood via CO_2. When the CO_2 tension of the blood is increased, respiratory receptors (in the *medulla, aortic arch* and *carotid sinus)* are *stimulated* by the gas itself, and by H ions of carbonic acid as a result, CO_2 gas is blown off, thus compensating for an acidemia. This explains why exercise increases our rate and depth of breathing. Con versely, when the blood is in a state of alkalemia there is a depression of respiration, causing a retention of CO_2 and H_2CO_3.

KIDNEY

The kidney plays a vital role in the regulation of the hydrogen ion. Renal failure can easily produce a fatal acidosis. The kidney functions in this capacity by protecting the buffer capacity of the blood, especially the Na phosphates. In acidemia, the Na_2HPO_4 (a basic salt) of the blood is used to aid in the neutralization of the H^+ ion. In other words, there is a shift in the type of phosphate salt, from Na_2HPO_4 the basic variety to $NaHoPO_4$ the acid variety. If this were to continue, the blood would reach an uncompensated acidemia. The kidney remedies the situation by *excreting* the acid salt into the urine, and *reabsorbing* the alkaline salt back into the blood. In this way the ratio of Na_2HPO_4 to NaH_2PO_4 is kept constant.

In addition, the kidney forms ammonia to replace the alkaline metals, Na and K, which have been used to neutralize acids in the blood. If this were not done, the blood would be deprived of its most valuable buffers: $NaHCO_3$; and $KHCO_3$. In alkalemia, the kidney acts in a reverse manner: It excretes the basic phosphate (Na_2HPO_4) instead of the acid phosphate (NaH_2PO_4), and Na and K salts instead of NH_4 salts.

THE ALIMENTARY TRACT

The cells of the stomach and intestine play an important role in regulating the *pH* by having the power of excreting acid

(HC1) and alkaline juices (containing $NaHCO_3$), respectively. In severe acidemia, for example, the blood's acidity stimulates the emetic centre directly, with the resulting loss of gastric HC1, thereby assisting in the restoration of pH. In cases of protracted emesis not due to acidemia, alkalemia may be produced. Likewise, diarrhea may cause an acidemia by a loss of alkaline intestinal juices. Excessive emesis and diarrhea, therefore, demand measures which will correct the shift in the blood's pH.

CLOTTING

Clotting prevents the escape of blood from the body. Intense investigations have disclosed that the clotting process is by no means simple; as a matter of fact, several theories have been offered to explain the series of chemical events involved. Most of these theories, however, agree on the fundamental steps. The following explanation is presented as a summary of contemporary thought.

When tissue cells and thrombocytes are injured (thrombocytes are injured by passing over a foreign surface), an enzyme called *thrombo-kinase* is released. Thrombokinase then acts upon a substance in the plasma, called *prothrombin,* to form an enzyme called *thrombin. This reaction, however, takes place only in the presence of calcium ion* (Ca^{4}+). Once thrombin is formed, it converts a soluble protein in the plasma, called *fibrinogen,* to *fibrin.* Fibrin separates from the blood as an *insoluble* mass of microscopic threads or fibers. In this way a meshwork is formed which catches and holds the cellular elements. This mass of fibrin and cells then contracts to form a *clot.* When exposed to air, the clot hardens and forms a tight seal (scab), thereby preventing the escape of blood. When blood is allowed to clot in a vessel, a clear aqueous liquid remains after the clot has formed; 'his liquid is called *serum.* Serum may, therefore, be defined as *plasma* minus *fibrinogen.*

CALCIUM

The role of calcium ion in the process is readily demonstrated by the fact that *sodium citrate* prevents clotting.

Sodium citrate reacts with calcium to form *insoluble calcium citrate;* in this way, Ca^{++} ions are *removed* from the plasma. Other salts, such as *oxalates,* will also prevent clotting by forming calcium precipitates. When blood is removed from the body for storage (in blood banks), a little sodium citrate is added to insure against clot formation.

PROTHROMBIN

Prothrombin is the inactive form of the enzyme thrombin. It is now known that prothrombin is synthesized in the liver, and further, that vitamin K is involved in the synthesis. Any injury to the liver, or anything which inhibits the utilization of vitamin K, will naturally inhibit clotting through a diminution of circulating prothrombin.

In medical practice, a drug called *Dicwnarol* is employed as an anticoagulant either therapeutically to prevent enlargement of *thrombi,* or prophylactically to prevent postoperative thrombosis. It is thought that Dicumarol acts by diminishing the rate of prothrombin synthesis in the liver.

THROMBIN

Thrombin is a protein enzyme catalyzing the conversion of *soluble* fibrinogen to *insoluble* fibrin. Another anticoagulant used in medicine, called *heparin,* is thought to act by either interfering with the formation of thrombin, or by inhibiting its action. The presence of heparin in the circulating blood and other inhibitory substances has indicated that the clotting process is not simple. For example, heparin, which inhibits thrombin, is in turn inhibited by another substance. As a matter of fact, the more that we learn about the biochemistry of clotting, the more complex the picture becomes.

FIBRINOGEN

Fibrinogen is one of the soluble protein constituents of plasma. It is formed in the liver and may be low (thus interfering with clotting) in subjects with severe hepatic damage.

THROMBOKINASE

Thrombokinase, or *thromboplastin,* is a phospho-lipid found in the thrombocytes (platelets) and tissue cells. When cells are damaged or when thrombocytes come into contact with a foreign surface (damaged tissues, glass, gauze, etc.), this enzyme is released. The cause of hemophilia has been suggested as being due to an *excess stability* of platelets.

OSMOTIC PRESSURE

The osmotic pressure of human cells, tissue fluid and blood plasma is equivalent to a 0.9 per cent solution of NaCl (such a solution is commonly referred to as *normal saline).* This pressure is due to *salts* (electrolytes), *soluble organic compounds* and *proteins.* It should be obvious that optimal cellular performance demands a constant pressure. The body, therefore, makes every effort to regulate the osmotic pressure of the blood and the various fluids derived from blood.

TONICITY

Solutions which have the same osmotic pressure are said to be *isotonic.* For example, blood plasma is isotonic with a 0.9 per cent solution of NaCl or a 5 per cent solution of glucose. For solutions which have different osmotic pressures *(anisotonic),* the terms *hyper-tonic* and *hypotonic* are used. Thus, a 10 per cent glucose solution would be relative to blood plasma *hypertonic;* a 2 per cent solution would be *hypotonic.*

When solutions are added to the blood stream, they should never be *hypotonic.* Hypotonic solutions cause red cells to swell and burst *(hemolysis).* Hemolysis is evidenced by the release of hemoglobin from within the cell, turning the solution red. Hypertonic solutions shrink *(crenate),* but may not destroy cells; thus, *small* amounts can be added to the blood stream. The relation of isotonic, hypotonic and hypertonic solutions to blood may be readily demonstrated in the laboratory.

ONCOTIC PRESSURE

Plasma proteins play a vital role because of the fact that they (normally) *do not* pass through the capillary wall. Hence, forced to remain in the blood stream, they tend to *hold* water; this affinity of protein for water is called *oncotic pressure.* Oncotic pressure is one of the determinant factors in the normal formation and reabsorption of tissue fluid: Tissue fluid is formed at the arteriolar end of the capillary where the hydrostatic pressure of the blood is *greater* than the oncotic pressure (22 mm. Hg); it is reabsorbed at the venular end where the hydrostatic pressure is *less* than the oncotic pressure.

Oncotic pressure, then, is one of the body's chief mechanisms for maintaining the volumes of the blood and tissue fluid. A loss of proteins from the blood, for example, causes a *loss* of blood volume and a *gain* of tissue volume (edema).

TRANSPORT OF CO_2 AND CK

The chemistry of the blood's physiology relative to the carrying of CO_2 and O_3 is an integral and inseparable part of the respiratory process. For this reason, the subject will be discussed under the respiratory system.

BLOOD FRACTIONS

Whole blood and plasma have been employed for quite a few years in replacement therapy. Very recently, synthetic substances (high molecular weight polymers) have been used with salutary results to restore blood volume following hemorrhage or shock. All of these methods, however, suffer important disadvantages; blood (especially whole blood) is not plentiful and, as yet, no substance has been synthesized which can replace blood. These facts provoked Dr. Cohn (a physical chemist) and his associates at Harvard to attack the problem in another way.

They reasoned that in most cases the use of whole blood or plasma is wasteful because certain *fractions* (e.g., albumin, globulin, etc.) of the blood are just as effective. Blood albumin, for example, is as effective in maintaining blood volume as plasma. One pint of blood, therefore, could serve the needs of more than one patient, each patient receiving the applicable fraction.

The fractionation of blood presented huge problems. This was the question: How can one separate a mixture of a hundred or more *similar* chemical entities (present in small amounts) without destroying their properties? Cohn and his workers supplied the answer by devising an extremely complex machine which completely processes the blood of the donor into cellular elements and plasma fractions.

The fundamental operation of the machine is *centrifugatlon;* i.e., the plasma fractions, after the cells have been removed, are separated much in the same way as cream is separated from milk when the latter is placed into a tube and spun at high speeds. At the last report, the machine has yielded about 70 fractions. The fractionation of blood is just another example of how the chemist has increased our span of life.

TISSUE FLUID AND LYMPH

Tissue fluid is a protein-low filtrate of blood plasma, produced by blood pressure forcing fluid through the capillary wall. This fluid forms the environment of the body's cells. Thus, all the essentials of life must pass through tissue fluid before permeating the cell's membrane. In addition, tissue fluid acts as a fluid reservoir in maintaining blood volume.

Lymph is the term applied to the excess tissue fluid which is removed from the intercellular area by so-called *lymphatic vessels.* Lymph may be considered a protein-low plasma, containing about one-half the amount of blood plasma protein. Lymph gathered from the *lacteals* during digestion, of course, contains fat. The small lymphatics empty their contents into the *thoracic duct* and *right lymphatic duct*. These are large valved vessels which connect with the subclavian veins, thereby permitting the lymph to aid in maintaining blood volume. Located along the lymphatics are *lymph nodes* which act as bacterial filters.

RESPIRATORY SYSTEM

Respiration includes the mechanism for inhaling and exhaling air, transport of CO_2 and O_2 by the blood, and the utilization of O_2 by the cells. When the blood passes through the lungs, CO_2 is given off and O_2 is absorbed; when blood passes through the capillaries of the tissues, the reverse occurs. It is important to appreciate that the chief force bringing about the *exchange* of O_2 and CO_2 is the *difference in pressures.* The actual *transport* of O_2 and CO_2 is via chemical mechanisms.The events are best understood by an analysis of the various reactions in the *lungs* and *tissues.*

LUNGS

The reactions in the pulmonary capillaries are the reverse of those in the tissue capillaries. These reactions are easily memorized if we start with the reaction between reduced hemoglobin of venous blood and O_2 of the pulmonary air. *oxyhemoglobin,* is an acid and reacts with $KHCO_3$ (also present in the red cell) to form *potassium oxyhemoglobin.*

Carbonic acid then decomposes (in the presence of *carbonic anhy-drase)* to CO_2 and H_2O, which are subsequently released from the red cell. The chloride shift continues until the blood's excess CO_2 is blown off. It is interesting to note that the red cell is *permeable* to HCO_3~ and Cl~ ions, but impermeable to K^+ and Na+ ions. If this were not so, all of the above reactions could not take place.

The air expired from the lungs is about 4 per cent CO_2 and 16 per cent O_2, as compared to atmospheric air of 21 per cent O_2 and a negligible amount of CO-. *Alveolar air* contains 5.5 per cent CO_2 and 14 per cent O_2.

CELLULAR RESPIRATION

The energy that a cell needs to maintain itself and perform its specific function or functions is supplied by the intracellular *oxidation of food.* Just how this energy is transformed into work has proved to be a vastly complex problem. In most instances little is known, but considerable progress has been made in elucidating the mechanism of muscular contraction. For the present, however, we are just interested in the way in which amino acids, glucose, fats and oxygen yield energy.

In the early years of physiology, it was thought that *intracellular,* or *biologic, oxidation* was more or less like oxidation in the test tube. We now know that this kind of oxidation involves a myriad of extremely complex *enzymatic* reactions. The following equation, therefore, is merely a *summary* of a long series of catabolic chemical events in which glucose is oxidized to CO_2, H_2O and *energy*.

$$C_6H_{12}O_6 + 6O_2 \rightarrow 6CO_2 + 6H_2O + \text{energy}$$

It is most important to realize that each step shown (and the ones not shown) requires a *specific enzyme and coenzyme*. Moreover, we should take special note of the fact that each coenzyme contains a *vitamin* from the *B complex group*.

Let us briefly describe the events. *Glucose* and *glycogen* react with phosphoric acid to yield *hexose phosphate*. (In muscle tissue, glucose must be in the form of glycogen; in other tissues, glucose may be used directly.)

Hexose phosphate then breaks down (passing through several intermediates) to *pyruvic acid*. Up to this point no oxygen has been used the so-called *anaerobic phase*. During this phase a certain amount of energy has been released at several of the stages. The next phase the *aerobic phase* involves the oxidation of pyruvic acid to CO_2, H_2O and *energy*.

This part of the process proved to be the most difficult to understand. The specific chemical events remained a mystery until Krebs (and others) worked out the details. For the sake of simplicity, we shall ignore these details and accept the fact that pyruvic acid is "burned" in the so-called *Krebs' cycle* to yield *waste products* and *energy*. There are, then, two sources of energy: that derived from the *anaerobic phase*, and that derived from the *aerobic phase*.

What about the metabolism of fats and amino acids? The answer to this question demonstrates the amazing powers of the biochemist. It is now known for certain that glycerin, fatty acids and amino acids are burned in the same "furnace" which consumes glucose; i.e., they are broken down and converted into the same *intermediates* which pass through the anaerobic and aerobic phases of glucose catabolism.

BASAL METABOLIC RATE

The sum total of all the energies required by the body's cells to *maintain* themselves is referred to as the *basal metabolic rate,* or *B.M.R.* A subject's B.M.R. is determined by measuring the amount of heat* eliminated from the body, and expressing this amount as *Calories (large) per square meter (of body surface) per hour.* During this test the subject must remain at *complete rest* (no physiologic or mental activity) to insure that we are reading only *basal energy.* Once the B.M.R. has been calculated, it is compared to *standard values* (derived from a large number of readings taken on normal subjects) and expressed as a per cent above or below normal. The basal metabolic rate is not considered abnormal unless it varies more than 10 per cent from normal. The taking of the B.M.R. is a valuable diagnostic medical test. The status of the thyroid, for example, is ascertained via the B.M.R. (Hypothyroids have low B.M.R.'s; hyper-thyroids have high B.M.R.'s.)

MUSCULAR SYSTEM

Energy for Contraction. Our chief chemical interest in muscle is concerned with the way in which muscle cells obtain their energy for contraction. Essentially this process is similar to that already presented under cellular respiration; it differs, however, in certain important aspects.

The contraction process requires a *sudden* jolt of energy. It has been appreciated for some time that energy of such a nature could not be supplied either from *glycolysis* (the breaking down of sugar or glyco-gen) or from the *oxidation of pyruvic acid.* This stimulated researchers in biochemistry to look for other plausible chemical mechanisms. As a result of these studies (especially those of Szent-Gyorgyi), the chemistry of muscle contraction is fairly well understood. The general idea of this complex picture may be described as follows.

The *Immediate energy* for contraction has been demonstrated to be supplied by the hydrolysis of *adenosine triphosphate* (ATP) to *adeno-sine diphosphate* (ADP) and H_3PO_4. The cell's supply of ATP, however, is small; ADP, therefore, must be reconverted into ATP. The energy for this reconversion, or synthesis, is supplied by the hydrolysis of *phosphocreatine* (PC). But the supply of phosphocreatine is also limited and, like ATP, must be re-formed. Once again, this requires energy; this time the energy is yielded by the *hydrolysis* of *glycogen to lactic acid.* Up to this point no oxygen has been used; i.e., the muscle cell has been able to contract in the *absence* of oxygen (the *anaerobic phase).* Of the lactic acid which has been formed, approximately 20 per cent is burned in the Krebs' cycle, yielding CO_2, H_2O and *energy*. This energy is then utilized to synthesize the remaining lactic acid (80 per cent) into glycogen. As can be easily perceived from the above steps, oxygen is the ultimate source of energy.

THE OXYGEN DEBT

Sustained muscle contraction piles up more lactic acid than the body can handle. Since this must be disposed of, we speak of the so-called "oxygen debt." This explains why the athlete continues to pant after the completion of his exercise *his body needs oxygen to burn up the excess lactic acid, and pay off its debt.*

MUSCLE FATIGUE

When excised muscle is electrically stimulated, the following happens: After the first few shocks the muscle seems to be more alert, and contracts with a greater force. Experimental evidence indicates that this is due to the accumulation of lactic acid, which in some manner sensitizes the muscle cells. Athletes take advantage of this chemical fact by mildly exercising "warming up" prior to the actual contest. Continuous and repeated shocking, however, places the muscle in a state of fatigue, a state in which the muscle is unable to respond to further stimuli. This results from a diminution of organic phosphates (ATP and phosphocreatine), and from the toxic effects of excess lactic acid on the muscle cells. Muscle fatigue in the body is probably due mainly to lactic acid formation, since the supply of organic phosphates is never completely exhausted.

TETANY

Muscle tissue, like other tissues, is extremely sensitive to a change in the *relative* concentrations of Na, K, Ca and Mg. In *tetany,* for example, we have a condition of extreme muscular contrac-tibility due to a diminution of Ca++ ions. That this is true can be demonstrated by the fact that injections of calcium salts readily reverse the condition.

NERVOUS SYSTEM

The manner in which a nerve fiber carries an impulse has always intrigued researchers in physics, chemistry and physiology. There have been two schools of thought on this subject the *electrical* and *chemical.* The electrical school holds that the passage of an impulse is analogous to the flow of current in a conductor; the chemical school likens the passage to a series of ultra-fast chemical reactions.

It now appears, however, that the ability of nervous tissue to carry an impulse is best explained by a synthesis of both theories; namely, that a nerve fiber is somewhat like a *battery.* Thus, we may think of a nerve impulse as a flow of *electrical energy* which results from chemical activity within the fiber's protoplasm.

The chemical activity, or metabolism, within nervous tissue exceeds that of other tissues. The brain, for example, has been estimated to contribute 10 per cent of the body's total basal metabolism. Because of this high metabolism, nervous tissue must receive a continuous and generous supply of glucose and O_2. This explains why a lack of either O_2 *(anoxia)* or glucose *(hypoglycemia)* easily damages certain vital centers in the brain.

Of considerable physiologic interest is the fact that the various types of nerve fibers have certain distinguishing chemical features. These chemical features are referable to the way in which an impulse behaves at *synapses* and *myoneural junctions.* On such a basis, there are two types of nerves: *cholinergic* and *adrenergic.*

CHOLINERGIC NERVES

Cholinergic nerves are rurve fibers which release *acetylcholine* at their endings. In this manner an impulse, after having arrived at the end of a nerve, is carried over to another nerve. Thus, we may look upon acetylcholine as a "physiologic electrolyte/'

Once the impulse has passed over the "gap" it is "erased" by an enzyme called *cholinesterase.*

ADRENERGIC NERVES

Adrenergic nerves are fibers which release an *epinephrine-like* substance called *sympathin* at their endings. Like acetylcholine, sympathin functions as a "physiologic electrolyte." Sympathin is hydrolyzed, or erased, by an enzyme located at the nerve ending.

ANATOMY VS. CHEMISTRY

For the sake of clarity, we should very briefly compare the anatomy of the nervous system with the chemistry presented above. Anatomically (and physiologically) there are two types of nerves *voluntary* and *involuntary,* or *autonomic.* The voluntary nerves activate *skeletal muscle,* and are under conscious control; involuntary, or autonomic, nerves activate *smooth muscle,* and are not under the control of our will. *All voluntary nerves liberate acetylcholine at synapses and myoneural junctions.*

Before we can make a similar statement for the autonomic nervous system, we must be familiar with the fact that an autonomic nerve has two parts a *preganglionic fiber* and a *postganglionic fiber;* furthermore, that there are two autonomic systems the *sympathetic* and the *para-sympathetic.* These two systems differ anatomically in respect to the *length* of their preganglionic fibers: Sympathetic nerves have *short* preganglionic fibers, whereas parasympathetic nerves have *long* ones. Chemically they differ in this respect: *Acetylcholine is formed at the ends of all preganglionic fibers and at the ends of postganglionic parasympathetic fibers, whereas sympathin is formed only at the ends of postganglionic sympathetic fibers.* Therefore, the following statement is true: All preganglionic fibers and all postganglionic parasympathetic fibers are *cholinergic;* all postganglionic sympathetic fibuis are *adrenergic.*

The above information has been extremely helpful to the physiologist and pharmacologist. For example, when acetylcholine is injected into the blood stream, it produces the same effects as those elicited by electrically stimulating the cholinergic nerves. Also, when *epinephrine* (Adrenalin) is injected, the results are referable to sympathetic stimulation. This, and a vast body of other facts, have made possible the development of a large number of useful drugs. There are four types of such drugs: (1) those that *mimic acetylcholine (parasympathomi-metics); (2)* those that *inhibit acetylcholine (parasympatholytics);* (3) those that *mimic epinephrine or sympathin (sympathomimetics);* and (4) those that *inhibit epinephrine or sympathin (sympatholytics).* These drugs are employed in those conditions which are benefited by either stimulation or inhibition of nervous impulses. *Atropine,* for example, helps the patient with peptic ulcer by inhibiting the parasympathetic fibers which cause the release of hydrochloric acid in the stomach.

THE ENDOCRINE SYSTEM

At various locations in the body there are small glands (the *endocrines)* which manufacture potent chemical regulators called *hormones.* Anatomically, the endocrines are distinguished from the *exocrine* glands (e.g., the salivary glands) by the absence of a duct; for this reason, the exocrines and endocrines are commonly referred to as *duct* and *ductless glands,* respectively. The endocrine glands are stimulated (or inhibited) to secrete their hormone or hormones by the nervous system, or by the hormones of other glands, often by both. The hormones thus produced enter the blood directly, and by some chemical mechanism regulate (by stimulation or inhibition) the physiology of some other tissue or tissues.

The isolation and synthesis of hormones by the chemist have made it possible for the physiologist to clarify many of his concepts of life's processes. In the field of medicine, hormones have enabled the physician to treat hopeless endocrine disturbances.

THYROID GLAND

The thyroid gland, located in the lower part of the neck, secretes an iodine-containing hormone called *thyroxin*. Thyroxin has the following chemical structure:

$$CH_2\text{-}CH\text{-}NH_2COOH$$

Thyroxin is a potent substance which controls the *rate of oxidative* processes in the tissues. *Hypothyroidism* (insufficient thyroxin) produces the clinical pictures of *cretinism* (in the child) and *myxedema* (in the adult). Both of the conditions are characterized by retarded growth of body and mind. The opposite condition, *hyperthyroidism,* is marked by nervousness, sleeplessness, loss of weight and muscular weakness. *Hyperthyroidism* (excessive thyroxin) results from the abnormal multiplication of thyroid cells *(hyperplasia).* This is evidenced by enlargement of the gland, producing a bulge in the neck. Such a condition is called a "toxic goiter." Toxic goiter should be distinguished from simple goiter. A simple goiter is caused by a lack of iodine in the diet, resulting in a decrease in the output of thyroxin. In order to compensate for this shortage, the gland enlarges, A simple goiter usually presents no problem other than the pressure caused by the enlargement.

Since thyroxin is concerned with the stimulation of intracellular oxidation, or *metabolism,* it is quite easy to understand that the *basal metabolic rate* (B.M.R.) is *decreased* in hypothyroidism and *increased* in hyperthyroidism. For this reason a B.M.R. test is a valuable diagnostic procedure.

Simple goiter may be *prevented* by the taking of extra iodine in the form of inorganic salts (Na, etc.). Such regions in the United States as the basin of the Great Lakes and the Pacific Northwest are referred to as *goiter regions* because of the lack of iodine in the soil and water. Until iodine salts were given (in drinking water and in table salt), simple goiter was prevalent in these areas. Toxic goiter, if severe, necessitates removal of part of the gland. In milder cases the so-called antithyroid drugs (e.g., *propylthiouracil)* are effective. By some mechanism, they interfere with the synthesis of thyroxin. Cretinism and myxedema are dramatically controlled by the administration of thyroid extract, which contains the hormone.

PARATHYROIDS

The parathyroids are four tiny glands embedded in the posterior aspect of the thyroid gland. These glands secrete a hormone called *parathyroid hormone.* Parathyroid hormone regulates the exchange of *calcium* between the bones and body fluids. *Hypoparathyroidism* (usually seen in those cases where the parathyroids were removed accidentally as a result of thyroidectomy) causes *tetany.*

Since a *reduction* in the calcium ion (Ca^{++}) of the blood (and tissue fluid) is the cause of tetany, we may conclude that parathyroid hormone in some way stimulates the bone to *release* calcium to the blood. Excessive parathyroid activity, or *hyperparathyroidism*$_y$ brings about *decalcifica-tion,* a condition marked by soft and fragile bones. Parathyroid hormone is employed clinically in those conditions which are ameliorated by increasing blood calcium.

PANCREAS

The pancreas is both a *duct* (secretes digestive juice) and *ductless* gland. The ductless portion, which is constituted by a specialized group of cells called the *islands of Langerhans,* manufactures one of the most important hormones in the body *insulin.* Insulin is essential for the *metabolism of carbohydrates:* It promotes the *synthesis* of glycogen *(glycogenesis),* and accelerates the *oxidation* of glucose in the tissues. These two facts have been known for many years.

The mechanism of insulin's function, however, has only recently been understood. Most authorities now agree that insulin acts by *facilitating* the *first step* in the synthesis of glycogen and oxidation of glucose. This first step is the formation of *hexose phosphate* from *glucose* and *adenosine triphosphate* (ATP).

DIABETES

Diabetes is an insidious and vicious disease which results from the inability of the islands of Langerhans to produce sufficient insulin for the body's needs. As a consequence, glucose accumulates in the blood *(hyperglycemia).* When hyperglycemic blood passes through the kidney, sugar appears in the urine *glucosuria.* For this reason, urine serves as a valuable diagnostic aid in the detection of diabetes.

The chief insult to the body arises from the abnormal metabolism of fat. Fats are burned because the body is unable to get its energy needs via the normal channel of glucose oxidation. In normal fat metabolism the liver converts fatty acids to simpler acids (e.g. *aceto-acetic* acid), which are then oxidized as an *alternate* source of energy. In diabetes, however, the liver is stimulated (by a *lack* of glycogen) to produce more aceto-acetic acid than the tissues can accommodate. As a result, aceto-acetic acid, together with its derivatives, *fi-hydroxybutyric acid* and *acetone,* accumulate in the blood. (These acid substances are called *acetone,* or *ketone, bodies.)*

If this continues, *acidosis* (referred to also as *ketosis)* results. It will be recalled from our earlier discussions that death results when the blood's pH drops below 7.2. The so-called *diabetic coma* is brought about by such a reduction in *pVL.* Ketone bodies in the blood "spill over" into urine. Their presence in the urine along with glucose is strongly suggestive of diabetes.

Diabetes is managed by limiting the carbohydrate in the diet, and by supplying insulin. In mild cases diet alone may remedy the condition; in severe diabetes, insulin is the only thing which will save the patient. The isolation of insulin by Banting and Best (1922) represents one of the most dramatic events in biochemistry.

ADRENALS

The two adrenal glands are located on the top of each kidney. Each gland has an outer portion called the *cortex,* and an inner portion called the *medulla.* The cortex secretes several hormones, whereas the medulla elaborates one hormone *epinephrine.*

PINEPHRINE

Epinephrine (also known as *Adrenalin)* was one of the earliest discovered hormones. Takamine, a Japanese biochemist, synthesized epinephrine in 1900. Epinephrine is the body's emergency hormone; i.e., all of its actions are designed to prepare the body for "fight or flight:" *raises blood pressure, stimulates the heart, causes glycogen to yield glucose (giycogenolysis), and stimulates the sympathetic nervous system.*

Epinephrine is one of the most potent and important drugs in medicine. It is especially valuable in shock, where it is used to increase blood pressure.

CORTICAL HORMONES

The adrenal cortex secretes several hormones which exert a profound influence on the body's metabolic processes; unlike epinephrine, they are essential for life. The role that these various hormones play in the chemistry of the tissues has proved to be a vastly complex subject. It is certain, however, that the hormones so far isolated influence either the *distribution of sodium and potassium, the metabolism of carbohydrates, or sexual development.* Thus, it is easy to understand that any cortical abnormality causes severe and widespread repercussions. *Addison's disease,* for example, a *hypocortical condition,* displays an array of signs and symptoms muscular weakness, reduced basal metabolism, reduced temperature, reduced blood volume and dehydration. A great many other pathologic disturbances are known to be caused by cortical malfunction.

CORTISONE

Of the several cortical hormones, one, for our purpose, deserves special mention *cortisone.* Cortisone, which was isolated from cortical extracts by Kendall (1949), has the so-called *steroid* structure, characteristic also of the other hormones of the adrenal cortex.

The advent of cortisone marks one of the great discoveries in the field of drug therapeutics. Once again the chemist has presented man with a magic cure via a magic formula. Although cortisone was originally introduced to relieve the extreme pain of *rheumatoid arthritis,* it has since been found to ameliorate 20 or 30 different clinical conditions. These diverse and miraculous effects can only be explained on a basis of its fundamental activity on the body's metabolic processes.

SEX GLAND HORMONES

The primary sex organs *testes* in the male and *ovaries* in the female secrete hormones which are necessary for the proper development of the accessory sex organs (male: prostate and other glands; female: mammary glands, uterus and vagina) and the so-called secondary sex characteristics such as voice, distribution of hair and shape of body. The testes secrete the hormone *testosterone;* the ovaries secrete principally *estradiol* and *progesterone.*

INTESTINE

It was stated during our study of digestion that digestive juice from the pancreas and bile from the gallbladder are released into the intestine. Since this happens only during the time food is present, there must be some mechanism which *controls* the flow of these two juices. It is now generally held that the hydrochloric acid in the material which leaves the stomach stimulates the intestinal mucosa to secrete two hormones *secretin* and *cholecystokinin.* Secretin is absorbed by the blood and carried to the pancreas, where it then stimulates the flow of pancreatic juices; cholecystokinin causes the gallbladder to empty its stored bile.

Other hormones concerned with digestion are *gastrin* and *entero-gastrone.* Gastrin, released by the stomach wall in the presence of protein, brings about the flow of gastric juice. Enterogastrone, released by the duodenal wall when fat is present in the chyme ejected from the stomach, quiets the stomach, thereby allowing sufficient time for the intestine to cope with its digestion of fat before receiving a new charge.

PITUITARY

The pituitary is a small gland located at the base of the brain. This gland is sometimes referred to and quite accurately so as the "master gland." The pituitary not only affects the physiology of a variety of tissues, but also directs the activity of the other ductless glands. Close examination will disclose that the pituitary is composed of two distinct parts the *anterior lobe* and *posterior lobe*. The anterior lobe synthesizes six or more hormones, and the posterior lobe, two.

EXCRETORY SYSTEM

The waste products from the cell are picked up by the blood as it circulates through the tissues. These wastes, in turn, are removed from the blood by the lungs (CO_2, H_2O) and by the kidneys (solids). The principal solid wastes include urea ($(NH_2)_2CO$) and the salts (viz., Na, K, Ca, Mg, NH_4) of HC1, H_2SO_4 and H_2PO_4. These substances (and a small quantity of other wastes) are dissolved in water and excreted as *urine*. Since a great many abnormal physiologic activities are reflected in the physical and chemical characteristics of urine, it serves as a valuable diagnostic device in medicine. The presence of sugar, for example, together with acetone, is strongly suggestive of *diabetes*. The table below presents the characteristics and values for normal urine.

GLUCOSE

There is normally a small amount of glucose in the urine, but not enough to give a positive test with Fehling's or Benedict's solution. When glucose is present in appreciable amounts *(glucosuria,* or *glycosuria)* it is indicative of *diabetes mellitus*. A temporary glucosuria is present following the ingestion of a high carbohydrate meal.

ALBUMIN

The presence of albumin in the urine *(albuminuria)* is indicative of impaired kidney function. Albumin may be detected by heating the upper layer of urine in a test tube. If a cloudiness appears, dilute acetic acid is then added to determine whether phosphates or albumin is present. Persistent cloudiness following the addition of acid (phosphates are soluble in acetic acid) is a positive test for albumin.

ACETONE BODIES

Acetone, or ketone, bodies include *acetone*, *(3-hydroxybutyric acid* and *aceto-acetic acid.* Normal urine contains only traces of these substances. In diabetes, however, they are present in sufficient quantities to produce a positive test. The test is performed by adding a mixture of *sodium nitroprusside* and *ammonium sulfate* to urine, and then overlaying with a small quantity of strong *ammonia water.* A positive test is indicated by the formation of a *pink to purple* colour at the junction of the two liquids.

BLOOD

Blood in the urine is referred to as *hematuria* if red cells are present, and *hemoglobinuria* if hemoglobin is present. Hematuria indicates a hemorrhage in either the kidneys or along the urinary tract; hemoglobinuria (due to the hemolysis of red cells) occurs in diseases such as scurvy and purpura. Very small amounts of blood can be readily detected by adding a solution of *benzidine* and *hydrogen peroxide* (the so-called *benzidine test)* to some urine in a test tube. The formation of a *blue colour* constitutes a positive test.

BILE

In jaundice the urine turns a *greenish-yellow* colour, due to the presence of bile pigments. There are several tests which have been devised to detect these pigments. One test (the Smith test) is performed by overlaying urine with tincture of iodine. A positive test is shown by the formation of an *emerald-green* ring at the point of contact of the two liquids.

PUS

A large number of pus cells in the urine *(pyuria)* are indicative of an infection of the urinary tract or kidneys. Their presence is detected by microscopic examination.

CASTS

Casts are living cells which appear in the urine in diseases of the kidney. Like pus cells, they can be readily seen by microscopic examination.

SKELETAL SYSTEM

Our bones contribute substantially to the total weight of the body. The bulk of bone (and teeth) is constituted by the elements calcium and phosphorus in the form of calcium phosphate. Of the calcium in the body, about 99 per cent is found in bone; of the body's phosphorus, 75 per cent is located there. From this information we can easily appreciate that the diet must contain relatively large amounts of foods with a high content of calcium and phosphorus.

The "laying down," or deposition, of bone requires not only the elements calcium and phosphorus, but also vitamin D. Vitamin D acts by promoting the absorption of calcium and phosphorus from the intestine, and by assisting in the splitting off of *phosphate ions* (from complex molecules), which are then free to unite with calcium ions.

As pointed out in our brief study of the parathyroids, the *parathyroid hormone* promotes decalcification of bone. It is by this mechanism that the body calls upon the calcium in the bone to supply (in time of need) the calcium ions for its various tissues. Muscle, it will be recalled, requires a definite concentration of the calcium ion. Parathyroid hormone and vitamin D, therefore, are physiologically antagonistic.

Chapter 10

Nuclear Chemistry

The discovery of x-rays by Rontgen in 1895 was the beginning of a new phase of man's life on earth the "atomic age." It is the purpose of this chapter to present briefly the basic ideas of atomic chemistry and its applications to our society. In contrast to our earlier studies, we are now interested in the chemistry of the atom's nucleus.

RADIOACTIVITY

The discovery of x-rays stimulated scientists to look for other sources of similar penetrating rays. The French scientist, Becque-rel, found that *uranium compounds* possessed this power. In 1903 Madame and Pierre Curie received the Nobel Prize for the isolation of a new element called *radium* an element two million times more potent than Uranium. At this time a new word appeared upon the horizon *radioactivity. Radioactivity is the term applied to all chemical substances which emit energy rays.* It was Rutherford and his associates who explained the nature of these mysterious radiations.

By the use of photographic plates and a powerful magnetic field they were able to demonstrate that all radioactive elements give off three types of rays. These rays were named *alpha, beta* and *gamma*. Alpha rays consist of *helium nuclei;* i.e., they are positively charged particles containing 2 protons and 2 neutrons; beta rays are streams of electrons derived from the disintegration of neutrons; gamma rays are similar to x-rays, but more penetrating. Of the three types, the gamma rays are the most powerful.

What is the explanation of radioactivity? The answer to this question is best supplied by stating that radioactive elements possess *unstable* nuclei.

In contrast to ordinary elements, the nuclei of radioactive elements are evidently unhappy about their *number of protons and neutrons;* i.e., they prefer some other combination than the one they have. In order to obtain the desired combination, the nuclei release alpha, beta and gamma rays. The radiation continues until the nuclei have attained their desired number.

HALF-LIFE

The disintegration of radioactive nuclei is spontaneous and not changed by outside forces. Each radioactive element, therefore, has a characteristic *rate* of disintegration called its "half-life." A radioactive element's half-life is defined as the time required for a given amount of the element to lose half of its mass. It would take 1590 years, for example, for 1 gm. of radium to reduce to 0.5 gm. Uranium has a half-life of several billion years. On the other hand, there are some radioactive elements which have a half-life of only a few minutes. When working with radioactive materials it is essential to know an element's half-life value.

TRANSMUTATION

As stated above, alpha and beta particles are shot out from the nucleus during the process of radiation. Since each element is composed of atoms whose nuclei contain a *characteristic number of protons and neutrons,* it follows that a radioactive element *ceases* to remain the same element. This means that a radioactive element can pass through several other elements before it reaches a stable form. The earth's *lead,* for example, resulted from the disintegration of *radium* which, in turn, resulted from the disintegration of *uranium*. This phenomenon of one element changing into another element is called *transmutation*.

The concept of transmutation served as the most powerful stimulus to the minds of the early atomic scientists. The implication was this: If natural transmutation (uranium to lead) is possible, why would it be impossible to bring about *artificial transmutation?* The great mind of Rutherford demonstrated to the world that artificial transmutation was possible. Rutherford succeeded in bringing about the transmutation of nitrogen to oxygen by knocking protons out of nitrogen nuclei, using alpha particles (supplied from radium) as "atomic bullets." Since this time (1919) a large number of other transmutations have been achieved, yielding new elements and *radioactive isotopes.*

RADIOACTIVE ISOTOPES

As previously defined, isotopes are atoms which have the *same atomic number,* but *different atomic weights.* Chemically, therefore, they are the *same.* It has been discovered, however, that an element's isotopes may vary in another respect radioactivity. Researchers have disclosed that many elements which heretofore were considered stable contain *traces* of radioactive isotopes. Various complicated physical and chemical techniques have been developed to isolate the radioactive isotope from the stable variety. It has been shown, for example, that ordinary carbon contains traces of a radioactive isotope with an atomic weight of 14 (written as $_6C^{14}$); the stable isotope of carbon has an atomic weight of 12.

THE DETECTION OF RADIOACTIVITY

As time goes on, more and more people are working with radioactive substances, it is essential, therefore, that we have suitable devices for the detection of invisible radiations. The simplest method is the use of *photographic plates.* Plates are exposed even when encased in metal shields. Other devices include the *electroscope, Wilson cloud chamber* and *Geiger-Muller counter.* At the present time the Geiger-Muller counter is used routinely in all establishments which employ radioactive elements and isotopes. The counter not only detects radiation, but also is able to furnish the *degree* of radiation.

ATOMIC MACHINES

The atomic scientist was not content with his knowledge of radioactivity and transmutation he wanted to probe still deeper into the atom's nucleus. How was it possible to learn more about the nucleus? Rutherford's work appeared to offer the greatest possibilities, i.e., the use of "atomic bullets." It was reasoned that once the nucleus was split, the resulting fragments would offer a wealth of atomic information. Rutherford's bullets *alpha particles* were not of sufficient velocity, however, to do the job; atomic bullets with unbelievable velocities were needed.

The answer came in 1929 when E. O. Lawrence invented the first "atom smasher" the *cyclotron*. By the use of huge electric currents and magnetic fields, the cyclotron imparted protons with an energy of 1.2 million electron volts. Since this time more powerful machines have been developed, such as the *betatron* and *synchrotron*. The latest atom smasher is reported to hurl subatomic particles at energies in the neighbourhood of *6 billion electron volts*.

The various atom smashers have enabled the physicist and chemist to study the nucleus in detail. When one reads of the advancements in this field it is difficult to believe that man possesses such fundamental knowledge. It is not true to assume, however, that matter has yielded its final secrets; indeed, the picture sometimes seems more complex. For example, it is now known that the proton and neutron are not the only inhabitants of the nucleus; scores of other particles (e.g., the *positron, meson, neutrino,* etc.) are present. So far, it remains to be seen just where all these subatomic particles fit into the architecture of the atom. In addition to the splitting of nuclei for the purpose of yielding basic information, the chemist and physicist have utilized their atomic machines to produce *artificial radioactive isotopes* and *new elements.* At the present time there are 101 elements 9 of which are *artificial*

NUCLEAR FISSION

In the early part of our study, we said that chemical reactions usually release energy sometimes in small amounts as in the rusting of iron, and sometimes in huge amounts as in the blast of TNT. Regardless of the magnitude, however, the energy is released by the same mechanism *the loss or gain of electrons.* In 1939 it was demonstrated that atoms are able (under certain conditions) to yield terrific amounts of energy by another mechanism *nuclear fission.* Nuclear fission is the term applied to the splitting of an atom's nucleus. Long before this experimental proof, it was realized that unbelievable amounts of energy were locked in the nuclei of atoms.

This theory was supported by the fact that if this tremendous energy were not present the nucleus would fly apart, since the protons of the nucleus, being of the same charge, would repel one another. It was further realized that if the nucleus could be split, this energy would be *released.* Until 1939, however, no one had been able to split atoms (in the atom smashers) in sufficient number to produce any significant amount of energy. Moreover, more energy was required to operate these giant machines than vvas released by fission.

The fission reaction of 1939 was concerned with an isotope of uranium (ordinary uranium is U^{238}). It was demonstrated that slow-moving neutrons penetrate the IP nuclei, forming U^{236}. Being unstable, IP^{35} immediately splits into smaller nuclei and two or more *additional neutrons.* These neutrons touch oil other U^{235} nuclei, so that the fission is not only sustained but *intensified.* This unique type of fission is referred to as a *chain reaction.*

The implication of this phenomenon was this: A pound or so of IP could yield energy with a force billions of times that possessed by an equivalent amount of TNT. We can perhaps better appreciate how the infinitely small nucleus is capable of releasing such energy if we consider Einstein's formula $E = mc^2$ (energy equals mass times the speed of light squared).

When the uranium nucleus splits, the sum of the nuclei which are produced weighs *less* than the original nucleus; matter, therefore, has been *converted into energy* in accordance with $E = mc^2$. What is important to appreciate is this: Even though the matter which disappears is unbelievably minute, it is being *multiplied* by the speed of light squared (186,000 miles per second). Moreover, this figure is further multiplied by the astronomic number of atoms present in a small amount of the material. The reason why a volcanic explosion did not take place in 1939 was because the reaction was *controlled;* that is, the bulk of the neutrons which were released by the fission of U^{235} were absorbed by inert material.

THE ATOMIC BOMB

As we all know, the *atomic bomb* came as an inevitable sequel to the chain reaction. The development of the bomb was the most complex and expensive project that man had ever undertaken. The chief obstacle resulted from the tremendous task of separating the infinitesimal quantities of U^{235} from uranium ore *constitutes less than 1 per cent of a given quantity of the element.* A huge chemical plant was built at Oak Ridge, Tennessee, exclusively for the production of uranium.

Soon after atomic scientists had started their work on U^{235}, it was discovered that another element possessed fissionable powers *plu-tonium.* Since plutonium could be produced from the relatively abundant U^{238} in a device called an *atomic pile,* there was no longer any question about the -success of the bomb. For the production and isolation of plutonium from uranium, another giant chemical plant was constructed at Hanford, Washington.

The bomb itself consists of two charges a small mass of U^{235} separated from a larger mass of plutonium. As long as these two fissionable masses are kept apart, no explosion occurs. This is because a good share of the neutrons which are being released by the fission of their nuclei are escaping into the air. When brought together, however, so as to exceed the so-called "critical mass," the neutrons are present in sufficient quantity to set off a chain reaction which is complete in about a millionth of a second. The two bombs used in World War II unleashed an energy equivalent to about 20,000 tons of TNT.

THE HYDROGEN BOMB

Already the atomic bomb has been made obsolete and tame in comparison to the hydrogen bomb. In contrast to the atomic bomb, the hydrogen bomb explodes by a *fusion* of its nuclei instead of fission. The principle, however, is the same mass is converted into energy. The hydrogen bomb is reported to contain a charge of *tritium* an isotope of hydrogen. When the tritium nuclei are detonated by a "small atomic bomb," they *fuse* to form larger nuclei (viz., helium).

A loss of mass results (the large nuclei weigh less than the sum of the smaller nuclei from which they were formed) and energy is released. Unlike the atomic bomb, there is no critical mass involved, which means that a hydrogen bomb may be made as *large as desired.* Since the sun derives its energy by the same mechanism, i.e., the conversion of hydrogen isotopes into helium, the chemist and physicist have duplicated one of the most fundamental reactions of the universe.

EFFECTS OF RADIATION

When an atomic device explodes, the blast represents only one of our fears the other is the radiation which is deadly to all living things within several square miles surrounding the blast. In addition, polluted air and water may be carried far into the interior of a region. The radiations accompanying an atomic blast are due principally to *gamma rays and neutrons.* Gamma rays are more penetrating and dangerous than x-rays. Unlike alpha and beta rays, which are stopped by thin sheets of metal, gamma rays are capable of passing through *several inches* of lead. Neutrons are dangerous to life in that they transmute certain of the body's elements (e.g., sodium and phosphorus) into radioactive isotopes. Although radiation sickness presents a concert of pathologic changes, the early effects are referable to the blood, where there is usually a reduction of both red cells *(anemia)* and white cells *(leukopenia).*

Animal studies and human studies (of those people who survived the atomic explosions in Japan) demonstrate that radiation can also bring about adverse inheritable changes. At present, however, this area of investigation is quite speculative. In spite of the terrible implications of radiation studies, there is a good side: What we learn about protecting ourselves can be applied to all those who handle radioactive materials for peaceful purposes.

ATOMIC ENERGY AND SOCIETY

Atomic energy is a term which has evolved to include all phases of nuclear chemistry *radioactivity, transmutation, fission, fusion,* and so on. When the time comes when man respects his fellow men, this energy will serve to enable society to construct a world far beyond our most vivid imaginations. Even though the threat of war has demanded that the scientists of the free world direct the bulk of their energies to the process of survival via nuclear weapons, there has still been a significant gain in peaceful pursuits. There seems to be no limit to the usefulness of atomic energy. Every area of science, medicine and industry has already enjoyed the fruits of atomic research. The following topics present the avenues through which the atom can serve all of us.

ATOMIC POWER

Atomic power is already a giant reality. The year 1954 will be remembered as the year of the first form of atomic locomotion (the submarine *Nautilus)* and the first atomic electric plant (located at Shippingport, Pennsylvania). Unlike conventional sources of power viz., coal, oil, water atomic power is without limit, and could be available to all people. If atomic power projects are permitted to flourish, every society will enhance its standard of living. There is no reason not to believe that atomic power will some day supplant all forms of conventional power.

The theory of atomic power is not difficult to understand. A small amount of fissionable material is placed into a special device, or holder, called a *reactor*. Since the fission is *controlled,* no explosion occurs; instead, great amounts of heat are released. The heat can then be used to heat water for steam, and the steam used to generate electricity. In practice, a reactor plan is not this simple; great problems had to be solved, such as finding a material which could withstand the heat of the reactor, and the disposal of radioactive wastes.

RADIOACTIVE ISOTOPES

So far, radioactive isotopes have been employed in innumerable ways in just about every branch of science. Their usefulness arises from the fact (already discussed) that they emit rays of energy.

THE TRACER TECHNIQUE

In the biologic sciences, radioactive isotopes are used to "tag" compounds in order to follow via the *Geiger-Muller counter* their disposition within plant and animal tissues. A tagged compound is prepared by replacing an element's stable isotope with a radioactive isotope. Carbon dioxide, for example has enabled the botanist to learn a great deal more about the process of photosynthesis. In general, this so-called *tracer technique* has brought forth information which otherwise could never have been elicited by conventional methods.

MEDICINE

The rays emitted by radium and radioactive isotopes have a profound influence on protoplasm. It is thought that the effects of radiation are due to the ability of the rays to *ionize* atoms and molecules, and in this manner alter the cell's chemistry. X-rays and radium have been used for years in the treatment of cancer. The availability of artificial radioactive isotopes, however, has enabled the physician to treat the patient in a more efficacious and economical way.

Moreover, certain isotopes offer special advantages. For example, radioactive iodine has proven to be highly effective in the management of cancerous thyroid. Radioactive iodine like the stable isotope accumulates in the thyroid tissue, and delivers a powerful dose of radiation at *exactly* the spot where it is needed. At the present time, several artificial isotopes are employed in medicine. The most frequently used include: carbon-14, sodium-24, phosphorus-32, cobalt-60, andiodine-131.

Chapter 11

Introduction with Inorganic Chemistry

Inorganic chemistry is not general chemistry. To some, this may appear as an obvious statement of fact. To others, it may appear as a contradiction of accepted opinion. As opposed to organic chemistry, inorganic chemistry is properly a study of all chemical materials other than the hydrocarbons and their derivatives. Not only does inorganic chemistry embrace the properties and modes of preparation of such materials, but it goes beyond and seeks to account for and explain specific characteristics and observed similarities, differences, and trends in observed behaviours.

Modern inorganic chemistry, unlike classical inorganic chemistry, is more than a descriptive science. It attempts to relate the properties of chemical substances to their structures, which in turn are related to the ultimate structures of the particles which combined to make those substances. Modern inorganic chemistry employs, therefore, both the experimental and theoretical approaches and requires familiarity with each. It is concerned with all the elements and with all the combinations in which these elements are found. Since it embraces a study of a variety of different individual materials, bonds, and behaviours, it is inherently more involved than organic chemistry.

Organic chemistry, though concerned with more individual compounds, involves a limited number of bond types and, therefore, a more restricted number of behaviour types. It is no exaggeration to state that the phenomena of organic chemistry will receive adequate theoretical interpretation long before those of inorganic chemistry.

Inorganic chemistry is not general chemistry. General chemistry is properly introductory chemistry and embraces only broad interpretations of the behaviours of chemical materials while providing the necessary bases for the further, more detailed studies involved in specific phases of chemistry. General chemistry amounts essentially to a combination of the elements of theoretical chemistry, analytical chemistry, organic chemistry, and inorganic chemistry. That many of the examples studied are inorganic in nature provides no real basis for terming it inorganic chemistry.

Emphasis on these materials is merely essential to an understanding of the periodic system as a whole. The unfortunate concept of synonymity has been maintained too long. It were well to discard it and then to develop the subject clearly and without preconceived views.

Classical inorganic chemistry was essentially purely descriptive. As such, it was the dominant phase of chemistry throughout the major portion of the nineteenth century. With the advent of ideas on atomic structure in the early years of the twentieth century, emphasis changed, and the resulting trend toward accounting for observed properties and behaviours rather than accepting and tabulating them as such has increased as more and more knowledge about the fundamental character of matter has been accumulated. This was the genesis of modern inorganic chemistry.

Its growth during the twentieth century has been largely along such lines, although necessary emphasis upon descriptive chemistry has been retained to provide data for systematic studies. The lines of demarcation between inorganic chemistry and other phases of chemistry have become less well defined. As a result, one finds in modern inorganic chemistry considerations of crystal structures, radioactivity, colour, analytical methods, oxidation-reduction phenomena, acid-base phenomena, physical measurements of a number of types, and a host of other things. To fit these into a logical picture which will then describe the scope of the entire subject is the primary purpose of the first part of this book.

Because so much of modern development in inorganic chemistry either involves atomic structure directly or is concerned with interpretations based upon atomic structures, it seems wise to base the general treatment of the subject upon the concepts of atomic structure. From these views, then, other concepts can be developed logically.

THE ATOMIC NATURE OF MATTER

The concept that matter is built up of tiny discrete particles probably originated with the early Greek philosophers. In fact, the designation *atom* for such a particle stems from the Greek meaning "not divided." However, it was not until about 1802-1803 that the rather nebulous ideas of the early thinkers were given the more quantitative interpretation necessary to modern development. At that time, John Dalton, the father of modern atomic theory, suggested that:

1. Matter is composed of tiny real particles called atoms.
2. Atoms of any pure substance can be neither subdivided nor changed one into another.
3. Atoms are incapable of being destroyed or created.
4. Atoms of any pure substance are identical with each other in weight, size, and other properties.
5. Atoms of one pure substance differ in weight and other characteristics from those of other substances.
6. Chemical combination amounts to the union of atoms in definite numerical proportions.

The fundamental importance of these thoughts cannot be minimized, even though some are no longer true in substance. Dalton's hypotheses were incapable of direct verification, but they were in accord with experimental fact and influenced chemical thought for over a century. It should be mentioned also that Dalton supplemented his general concepts with values for atomic sizes and weights. These values are without real significance.

Developments subsequent to Dalton's proposals amounted to expansions of the atomic theory and more comprehensive explanations of its tenets. Thus Prout (in 1815) proposed that the weights of all atoms were simple multiples of the weight of the hydrogen atom and that hydrogen was thus the fundamental material out of which all other elements were constructed. Although fundamentally incorrect. Prout's hypothesis does contain the ideas that atoms themselves are complex and that certain even more simple particles may exist.

As the nineteenth century progressed, this thought grew in stature, particularly after the work of Michael Faraday (in 1833) showed matter to be electrical in nature and demonstrated equivalencys between electrical energy and chemical change.

The complexity of the atom became increasingly apparent with discovery of the electron, of x-radiation and other types of radiation, of radioactivity, of nuclear reactions, and of a number of subatomic particles. The inescapable conclusion based upon the wealth of accumulated data is that a number of *particles* are essential to the construction of the atom. These particles may now be considered in detail.

FUNDAMENTAL COMPONENTS OF ATOMS

It is convenient to consider atoms as being composed of several types of ultimate particles. Some of these are of sufficient stability to be capable of independent existence outside an atom. Others are so unstable by comparison as to exist only momentarily outside the atom or to possess only transient existence within the atom itself.

Most of these particles are simple in nature, although some composite particles are known as well. However, there is little or no positive evidence that any composite particle plays a real role in the construction of an atom. Only the salient features of these particles need be considered here. For more detailed consideration, reference should be made to such a comprehensive text as that by Stranathan.

Stable Particles

Only three such particles are recognized, namely, the negative electron, the proton, and the neutron.

THE NEGATIVE ELECTRON

The term electron was first employed in 1874 by Stoney, who used it for the unit charge on a monovalent negative ion. Because of errors, however, the currently accepted magnitude of this charge is some sixteen times Stoney's value. In 1879, Sir William Crookes showed that in highly evacuated discharge tubes so-called cathode rays streamed from the negative poles when high potentials were applied. Regardless of the natures of the negative poles used or the natures of the residual gases in such tubes, these cathode rays were found to possess the following characteristics:

1. Travel in straight lines from the negative pole (cathode).
2. Cast shadows when targets are placed in their paths.
3. Produce mechanical motion in pinwheels.
4. Produce fluorescence in glass walls of the tubes.
5. Heat thin metal foils to incandescence.
6. Impart negative charges to objects in their paths.
7. Suffer deflection in applied electrostatic or magnetic fields.

Later, it was also shown that they cause ionization in gases, expose photographic plates, and yield penetrating x-radiation when directed against suitable targets. These characteristics suggest that such rays are made up of small but energetic negative particles. This was proved in 1897 by Sir J. J. Thomson. These particles are called *electrons*.

Although independent evidences for the existence of electrons in matter were obtained through the production of identical particles by thermionic and photoelectric emission and as a result of radioactive decay, perhaps the most significant studies were made with cathode rays.

Using tubes in which a collimated beam of cathode rays could be deflected by the simultaneous application of magnetic and electrostatic fields arranged perpendicular to each other, Thomson evaluated the ratio of charge to mass *(e:m)* for the electron by measuring the magnitudes of the two fields required to produce no deflection in the electron beam.

Although the value obtained by Thomson has been revised upward by more precise measurement, his observation that the ratio e:m is the same regardless of the nature of the cathode or residual gas has been shown to be valid. This leads to the conclusion that all electrons are identical. Evaluation of the electronic" change was first effected by Millikan by use of the classic oil-drop experiment, in which the rate of fall of a tiny charged oil droplet due to gravity was exactly balanced by an electrostatic field of measured intensity.

Millikan's value was accepted as exact for many years, and only recently have more precise measurements necessitated revision of it. The numerical characteristics represent the most generally accepted values at the present time. The reader will understand that literally hundreds of experiments in addition to those described have been involve^ in the complete characterization of the electron.

It should be pointed out that, although the electron has been characterized as a finite particle and is most usefully so considered by the chemist, it does possess wave properties. Although it is beyond the scope of this book to explore this view, such properties as reflection and diffraction can best be explained by means of such considerations.

These views on the wave properties of an apparently particulate material were first advanced by de Broglie and have since formed the basis for theoretical views on the extranuclear structures of the atoms. Their treatment lies properly in the realm of theoretical physics.

THE PROTON

Detailed experiments with discharge tubes of the cathode ray type showed the presence of beams of positive particles (positive rays or canal rays), which unlike the electrons of cathode rays depended for their charges and masses upon the natures of the residual gases in the tubes. However, it was found that unipositive particles with maximum ratio of charge to mass were obtained when the residual gas was hydrogen. These particles, called *protons,* were found to be identical with hydrogen atoms from which the single electrons had been removed. The appearance of protons as products in a variety of nuclear reactions has led to the general feeling that their presence in all atomic species is likely.

THE NEUTRON

Existence of a neutral particle common to most atomic species was predicted around 1920, but the difficulties associated with detecting a particle which could neither ionize other materials except by direct collision nor be deflected by magnetic or electrostatic fields were so great that proof of its existence was long in coming. The discovery of the neutron was foreshadowed by an observation by Bothe and Becker that bombardment of certain of the lighter elements (e.g., Li, Be, B) with alpha particles yielded a penetrating radiation, an observation which was checked by others but attributed to gamma radiation.

In 1932, Chadwick showed definitely that such radiation was due to neutral particles of approximately unit mass generated by nuclear reactions. Although similar reactions are difficult to effect with the heavier elements, evidences from a variety of nuclear reactions, especially those of the fission type support the view of the nearly universal occurrence of neutrons. Outside nuclei, neutrons appear to be unstable.

Unstable Particles

Among such particles are the positron, the neutrino and antineutrino, and the meson.

THE POSITIVE ELECTRON OR POSITRON

The positive counterpart of the electron was discovered more or less accidentally by Anderson in 1932, when, during studies on the effects of magnetic fields upon particles ejected from atomic nuclei by absorption of cosmic rays, he noticed fog tracks exactly like those given by electrons but deflected in the opposite direction. Particles responsible for these tracks were called *positrons.* They apparently resulted from cosmic radiation through the creation of electron-positron pairs. They are also noted in the decay of certain radioactive substances.However, their stable existence under any circumstance appears highly unlikely since electrons and positrons mutually annihilate each other with the production of gamma radiation.

THE NEUTRINO AND ANTI-NEUTRINO

These are particles of small mass and zero charge the existence of which has been postulated to account for energy changes during the radioactive emission of electrons and positrons. The neutrino is supposedly associated with and shares the energy of the electron, whereas the anti-neutrino occupies the same position with respect to the positron. Although these particles seem very probable, no concrete evidence for the existence of either has been obtained.

THE MESON (FORMERLY MESOTRON)

To account for nuclear binding energies, Yukawa postulated the existence of a particle (the meson) of mass intermediate between the electron and the proton. Mesons were discovered first in cosmic ray studies, resulting doubtless from the effects of cosmic radiation on matter. Recent work in ultrahigh energy ranges with the cyclotron has led to the creation of mesons in the laboratory.

Composite Particles

The only composite particles which are of importance are the deuteron and the alpha particle.

Chapter 12

Atomic Nuclei and Properties

Some detailed considerations on atomic structure may now be offered. Of the two general regions into which the structure of the atom has been divided, that involving the nucleus will be discussed first since it bears less relationship to the majority of the topics to be considered in this book than does the extranuclear region. Although it may be argued that discussions of atomic nuclei lie more properly in the fields of applied and theoretical physics, the line of demarcation between physics and chemistry has become so indistinct in recent years that neither of these branches of physical science can any longer be considered independent of the other. If the chemist is to employ the outer architecture of the atom to best advantage in solving his problems, he must be conversant with the inner architecture as well and with those properties which are dependent upon it.

Although knowledge relative to atomic nuclei has increased tremendously in the past few years, it must be admitted that surprisingly little is known in comparison with what remains to be learned. This is particularly true of theoretical considerations. Much of what is offered in a theoretical way to account for the constituents of atoms and their behaviours amounts more exactly to restatement of observed fact and cannot be used for further predictions. This is due largely to the lack of adherence of particles of nuclear and subnuclear dimensions to classical behaviour. However, much useful information concerning the behaviours of nuclei, their syntheses, their stabilities, the release of energy from them, and their general properties can be presented. If such fundamental points as the exact compositions of atomic nuclei and the true natures of the binding forces which render them stable remain unsettled, they should be regarded as goals for future achievement.

In the discussion which follows, no attempt has been made to cover more than the real essentials. The treatment is non-mathematical and, where possible, modular in character. More comprehensive treatments as indicated in the cited references may be consulted with profit by the interested individual.

THE COMPOSITIONS OF ATOMIC NUCLEI

A number of combinations of the fundamental particles discussed in Chapter 1 lead to nuclei with the requisite mass numbers *(A)* and positive charges *(Z)*. Among the more important nuclear pictures which have been offered are those which are described here.

THE PROTON-ELECTRON CONCEPT

Offered shortly after development of the nuclear atom concept, this view, in spite of its weaknesses, was quite generally held until the discovery of the neutron in 1932. The nucleus was considered to contain sufficient protons *(A)* to account for its mass and sufficient electrons $(A - Z)$ to neutralize the proton charges in excess of those indicated by the atomic number *(Z)*.

Both nuclear protons and electrons were regarded as essentially free, and a distinction between extranuclear and nuclear electrons was always necessary. The improbability of the coexistence of oppositely charged particles within the small element of space (sphere of diameter 10^{-12} cm.) encompassed by the atomic nucleus caused many to question the proton-electron concept and led both Harkins and Rutherford to postulate the existence of the neutron as early as 1919. The improbability of such a charge combination must be regarded as the strongest argument in support of the ultimate rejection of this view.

THE PROTON-NEUTRON CONCEPT

Discovery of the neutron was followed logically by an alteration of the nuclear picture to accomodate this neutral particle. Since the neutron was regarded as a close combination of an electron with a proton, the nucleus was considered to contain protons equal in number to the atomic number (Z) and neutrons *(N)* equal in number to the mass number *(A)* minus the atomic number. The proton-neutron concept is the most consistent in accounting for the majority of the properties of atomic nuclei and is the most widely held at the present time. Nevertheless, it is far from perfect. It is difficult, for example, to conceive of the stable packing together of numbers of positive particles in the small element of volume that is the nucleus. Nor in the usual sense can one conceive of sizable attractive forces between neutrons and protons. Although it is perhaps wrong to visualize the nucleus as a mere grouping of neutrons and protons, such a view is qualitatively useful. To regard a neutron as being a combination of a proton and an electron is probably equally in error, but it is again of qualitative use.

ISOTOPES

Of the postulates of Dalton's atomic theory, only the one suggesting that all atoms of a given element are identical to one another and equal in weight has been proved to be seriously in error. Actually, f identity is taken to mean identity in chemical characteristics, only the idea of equality in weight need be seriously altered. Such a revelation although incapable of proof in Dalton's time, seemed so logical that even the many fractional atomic weights found during attempts to prove Front's hypothesis failed to shake faith in it that fractional atomic weights necessitate the assumption that atoms themselves must have fractional weights for acceptance of this base of Dalton's theory apparently troubled few, if any, investigators.

This general concept was first questioned in 1886 by Sir William Crookes.The following quotation from Crookes's original publica-ion summarizes his views: "I conceive that when we say the atomic weight of, for instance, calcium is 40, we really express the fact that, while the majority of calcium atoms have an actual weight of 40, there are not a few which are represented by 39 or 41, a less number by 38 or 42, and so on." When one realizes that this statement was based upon no physical or chemical evidence, one is struck by Crookes's Imost prophetic insight into the nature of matter. Yet this concept /as not destined for acceptance for some time.

Shortly after this proposal, Crookes separated the earth into a number of components with similar chemical properties but different phosphor-iscence spectra. These he regarded as containing the same element but with differing atomic weights. Subsequent work by others showed his postulated *meta-elements* to be different elements, and the concept of variability in atomic weight was abandoned.

More conclusive evidence of the truth of Crookes's early speculations was presented shortly after the discovery of radioactivity. In 1906, Boltwood discovered a new radioactive element, which he found to have chemical properties so similar to those of thorium as to render its compounds chemically inseparable from those of thorium. Further work by others showed differences in mass and radioactive properties which proved ionium and thorium to be different materials. Yet even the arc spectra of the two were identical. The chemical identities of mesothorium-I with radium and of radium-D with lead were established at about the same time, and Soddy suggested that "chemical homogeneity is no longer a guarantee that any supposed element is not a mixture of several different atomic weights, or that any atomic weight is not merely a mean number."

Final evidence was presented in 1913, when it was shown by positive ray analysis that neon contains atoms of mass numbers 20 and 22, those of the latter type constituting much the smaller fraction of the whole. At the same time, Soddy, in considering the radioactive elements, suggested the term *isotope* (Greek, equal places) in the following words: "The same algebraic sum of the positive and negative charges in the nucleus when the arithmetical sum is different gives what I call 'isotopes' or 'isotopic elements' because they occupy the same place in the periodic table. They are chemically identical, and save only as regards the relatively few physical properties which depend upon atomic mass directly, physically identical also."

Subsequent investigations have shown that the occurrence of elements as mixtures of isotopes is the rule rather than the exception. Isotopes may be defined as atoms the nuclei of which contain the same number of protons but different numbers of neutrons. It follows, therefore, that all isotopes of a given species contain the same number of electrons, and, since these electrons are identically arranged, it is not surprising that essential identity in chemical characteristics should result. Such differences in properties as are due to mass differences can be large only with the lightest elements because it is only with them that percentage differences in mass are appreciable.

THE DETECTION AND STUDY OF ISOTOPES

One of the most important experimental approaches to the study of isotopes has involved positive ray analysis. Although discovered in 1886 by Goldstein as luminous paths behind a perforated cathode in a discharge tube, positive rays were not examined systematically until much later, when their production was shown to be a common phenomenon and their natures to depend upon the type of residual gas present in the tube.

Like electrons, positive rays are deflected by electrostatic and magnetic fields, and the extents of such deflections in fields of known strength may be used for the evaluation of charge to mass *(e:m)* ratios. The parabola method devised by Thomson was the first used for accurate studies. In this procedure, a narrow beam of positive particles, obtained by passage through a small hole in the cathode of a discharge tube, was passed through electrostatic and magnetic fields of known strengths applied simultaneously.

Deflections were measured by recording the position of the emergent beam upon a photographic plate. It can be shown that under such conditions all particles of the same charge to mass ratio but of differing velocities will fall along a parabolic path recorded on the film. In an actual experiment, a variety of parabolic paths, each characteristic of a particular positive ion species, is obtained. From the constants of a given parabola and the known field strengths, the characteristic charge to mass ratio can be evaluated.

Exact interpretation of experimentally determined positive ray parabolas is complicated by a number of factors. Thus every ionizable material in the discharge tube will yield at least one parabola, and careful account of the purity of the sample must be taken. Furthermore, a single material will commonly yield more than a single ionization product. For example, an analysis of carbon dioxide gave parabolas corresponding to the species C^+, O+, and $O+^2$. Each parabola requires careful characterization.

In spite of these difficulties, positive ray parabolas may be used to show the existence of isotopes and to evaluate their mass numbers. In 1913, Thomson first ascribed particles of mass 22 found in the analysis of neon to an element of similar properties occurring in the neighbourhood of neon, but these observations were soon shown to be due to a neon isotope of this mass number.

Nearly simultaneously, both Dempster and Aston adapted positive ray procedures to the study of mass spectra. Both procedures differed from Thomson's essentially in the application of the magnetic field perpendicular to the electric field in order to effect focusing of all particles of a given charge to mass ratio. Positive ions from a heated filament *F* were accelerated by an electric field applied between this source and a slit Si, and then bent by a magnetic field *H* applied perpendicular to the plane of the drawing.

Those passing through a second slit *S* were then measured as a positive ion current with an electroscope at *E*. Since it can be shown that the masses of the positive ions reaching the electroscope are dependent upon the accelerating potential and since the measured positive ion current is proportional to the number of positive ions striking the electroscope, Dempster's procedure amounted to varying the accelerating potential and recording a positive ion current at each potential. Plots of positive ion current against mass number then yielded peaks indicating isotopes at definite mass numbers, the heights of these peaks being proportional to the relative abundances of the isotopes. Dempster's apparatus used the focusing action of the -magnetic field to bring together particles having the same velocity but differing in direction. Aston's apparatus, however, brought together at one spot all positive ions having the same direction and the same charge to mass ratio, regardless of their velocities.

A narrow bundle of positive rays admitted through a hole in the cathode *C* was collimated by the slit system *S1* and S2 and passed between charged plates PI and P2, PI being positively charged. The resulting angles of deflection were dependent upon the ion velocities, the slower ions being deflected through greater angles than the faster ones. After passage through a third slit S_3 the divergent beam was passed through a magnetic field so applied as to produce deflections opposite to those obtained in the electric field. Ultimately, all particles of the same *e: m* ratio were brought to a focus at a given point on the photographic plate *P*. This apparatus is called a mass spectrograph.

From the positions of the exposed regions on the developed photographic plate, corresponding *e:m* ratios were evaluated. Actually, Aston was concerned more with the individual masses of the particles themselves. For an element yielding ions of the same charge, exposed regions would indicate the masses of the isotopes if the same reference standard was used. Aston referred these positions to that of oxygen, the mass of which he took as 16.0000, and obtained isotopic masses accordingly. Relative isotopic abundances were evaluated from the intensities of the lines on the developed plates.

Positive ray analysis has been extended and refined extensively, particularly by Aston, Dempster, Bainbridge, Jordan, Mattauch, and Nier, to mention several of the more prominent investigators. It is beyond the scope of this treatment to discuss in detail all the instruments employed and the results obtained. Reference to any of a number of excellent reviews is recommended. Instruments used may be described as either mass spectrometers or mass spectro-graphs. Mass spectrometers are devices which measure the intensities of positive ion beams, usually electrically, and which are employed to establish the abundances of the isotopes. These may be classified as radian spectrometers, magnetic lens spectrometers, crossed-field spectrometers, and sector spectrometers depending upon the mode of operation. Mass spectrographs, on the other hand, are devices which correlate the positions of positive ion beams, usually photographically, with ion masses and which are used to evaluate the masses of isotopes. Mass spectra are being used for ordinary analytical purposes increasingly often.

For the detection of comparatively rare isotopes, the optical method is sometimes more suitable than the positive ray procedure. The optical method depends upon the fact that the positions of rotational and vibrational lines in the far and near infrared regions of emission spectra, respectively, are influenced by the masses of the species which produce them. Slight line displacements, therefore, characterize the isotopes of a given element, and the extents of these displacements may be used in evaluating isotopic masses. The existence of both the heavier oxygen isotopes and the heavy hydrogen isotope, deuterium was indicated by optical data.

GENERALIZATIONS RELATIVE TO ATOMIC NUCLEI

In the range of nuclear charges from 0 through 83, stable isotopes of all mass numbers except 5 and 8 are known. Some 274 stable isotopes have been recognized, and of the elements listed only 23 are simple (i.e., non-isotopic or pure elements). Nuclei of the same mass number but differing charge, the so-called *isobars,* are not particularly uncommon; but, in a given set of isobars, only one member is normally abundant. The only exception is that of the $A = 40$ isotopes of argon and calcium, both of which are abundant. Such a condition is apparently one of instability or its incidence would be more common.

In addition to the unstable nuclei characterizing all nuclear charges above 83, similar species are known for all other charges. In total, nuclei with charges from 0 through 98 and mass numbers from 1 through at least 246 have been characterized.

Other generalizations may be summarized in terms of the so-called rules of Harkins. These "rules" appeared originally as a series of periodic properties listed by Harkins in a treatment of the proton-electron nucleus, but in a slightly different form they are equally applicable to the proton-neutron nucleus. Modified, these rules are:

1. *No common nucleus except hydrogen contains fewer neutrons than protons.* No stable nucleus except those of H and He has a mass less than twice its charge. Among the more common and more stable species, the mass number is commonly twice the charge, indicating a tendency toward equality in the numbers of neutrons and protons. Neutron to proton ratios never exceed 1.6, and those nuclei with neutron to proton ratios exceeding 1.2 are very uncommon in nature.

2. *Elements with even nuclear charges are more abundant, more stable, and richer in isotopes than those with odd charges.* Data indicate that elements with odd nuclear charges never have more than two stable isotopes and that elements with even nuclear charges are uniformly richer in isotopes than their odd-numbered neighbours. All stable elements having odd charges consist primarily of one nuclide. Among the lighter elements, those with even charges are often more abundant than their odd-charged neighbours. All those elements the existence of which in nature has

been questioned have odd charges, e.g., 43, 61. The half-lives of radioactive isotopes indicate the even-any symbol X, this notation becomes numbered materials to be the more generally stable. Alternation between odd- and even-charged nuclei is very apparent.

3. *Nuclei with even numbers of neutrons are more abundant and more stable than those with odd numbers.* Inasmuch as the majority of nuclei with odd charges have odd mass numbers whereas those with even charges have even mass numbers, it is apparent that the number of neutrons in a nucleus tends to be even. The most abundant nuclei in general contain even numbers of neutrons, the only exceptions being Be and N. Only tin has more than two stable isotopes with odd mass numbers.
4. *Nuclei with even mass numbers are more abundant than those with odd mass numbers.* For a given element, isotopes of even mass numbers are commonly more abundant than those of odd mass numbers. All the stable pure elements listed have odd mass numbers as well as odd nuclear charges. That the mass numbers of many of the most abundant and stable isotopes are multiples of four suggested originally that the alpha particle was a nuclear component. It is more probably the result of neutron-proton pairing. Some 87% of the nuclei in the crust of the earth have even mass numbers.

Nuclei are commonly classified in terms of whether they contain even or odd numbers of protons *(Z)* and neutrons *(N)*. Four such combinations are possible. The most common arrangement is *even-even* and the least common *odd-odd*.

13

Basics of Physical Chemistry

PRINCIPLES OF THERMODYNAMICS

The science of thermodynamics deals with the laws and relationships which govern the quantitative transformations of *energy* of one sort into energy of some other sort during physical or chemical changes in a system. The name suggests the mutual relation of heat and motion, but the principles involved have a much larger scope, and may be applied under given conditions to all forms of energy. The characteristic feature of thermodynamics is that it permits us to deal with energy changes involved in a physical change of state, or in a chemical reaction without in any way requiring information regarding the *molecular mechanism* of the process under investigation.

The conclusions which we shall arrive at on the basis of thermodynamics are thus independent of any molecular hypothesis we may have formed in respect of the process. This, it will be seen, is in many ways a very great advantage. It means that the conclusions of thermodynamics are quite general, and will remain true even if our views regarding the actual mechanism of the process considered from a molecular standpoint were to undergo radical change. Of course it will be seen as the converse of this that thermodynamical reasoning and conclusions, although true and general, do not tell us anything of the mechanism involved in a given process. This is naturally a considerable drawback, for advances in theoretical treatment seem to be most easily made along mechanical lines of thought, *i.e.* with the aid of molecular hypotheses. At the present time, therefore, the line of attack upon any problem which promises to be most successful is, in general, that in which we make simultaneous use of both generalised principles (thermodynamics) and the specialised principles (such as the Kinetic Theory).

The classical theory of thermodynamics may be said to rest upon two main laws, or fundamental principles, known as the First and Second Laws respectively. Recently another principle has been brought forward by the noted German scientist, Professor Nernst, of Berlin, which seems likely to ultimately take its place as a third law. This principle and its applications are discussed at the conclusion of our study of affinity.

THE FIRST LAW OR THERMODYNAMICS

This law is simply a statement of the principle known as the Conservation of Energy, according to which energy can be changed from one form to another, but can never be destroyed. Before discussing this let us consider briefly what we mean by the term " energy ".

When a body or system can do *work* against a the Force it is said to possess energy. How are we to obtain a *measure* of this energy? This is effected by regarding the decrease in energy as equal to the work done. When the energy arises from motion—say of the molecules of a system, or of the system as a whole in space—it is called Kinetic Energy. When it arises from the relative position of bodies it is called Potential Energy.

From the standpoint of thermodynamics, however, we do not distinguish the *origin* of energy whether it be kinetic or potential (as this would really involve molecular hypotheses); instead, we deal simply with the energy possessed by a body, or more frequently with the energy gained or lost by a system when the system undergoes a given change, without specifying more closely whether the energy thus lost or gained is kinetic or potential in nature.

It is important to notice that although we can mentally conceive of the idea of a given substance or system possessing energy, we are unable to give this a *numerical* value. What we can do, however, is to ascribe a numerical value to the *change* in energy involved in a given process; and it is with such *changes of energy, i.e.* either gain or loss of energy, that thermodynamics deals. A simple illustration of a mechanical nature will make the distinction clear. Suppose a body having the mass of i gram falls a distance of meter towards the earth under the action of gravity.

The *loss* of potential energy sustained by the body in thus altering its position is gram-meter, this being the work done in the change considered. We do not know the absolute value of the potential energy of the body in its initial position, but we do know that in its final position it has lost an amount, gram-meter. Now, the study of mechanics has made us familiar with the idea that the term " work," or " energy," is a composite one, being always expressible as the product of two terms, one -term being known as the " intensity factor," the other as the " capacity factor ". The simplest conception of work or energy is to regard it as the product of force into distance. The force is the " intensity factor," the distance the " capacity factor ". A force has the dimensions and a distance the dimension L, so that energy or work must have the dimensions.

Even when the actual energy is not simply force x distance it must still have the dimensions above ascribed to it Thus let us consider another sort of simple process in which work is done and the equivalent energy expended, namely, the isothermal expansion of a gas against the atmosphere or other pressure. Suppose the initial volume of the gas was v c.c., and the increase in volume is a very small one dv. Suppose p is the atmospheric pressure (in dynes per square centimeter). Let us suppose the gas is in a cylinder fitted with a weightless, friction less, movable piston. The surface area of the piston may be A cm.^{2}. In the equilibrium state the pressure above and below the piston must be identical. Under these conditions the piston would remain motionless for an infinite time. Suppose, however, that the gas pressure, *i.e.* the pressure exerted upon the lower side of the piston, is something greater than 1.

The gas will tend to push the piston outwards. Suppose it does so through an infinitely small distance *dx*. The work done = force x distance = *(p* x A) x *dx,* but *hdx* =» *dv;* .-. work = *pdv.* Now, in performing this expansion against the external pressure /, the gas has done work upon the surrounding atmosphere, or, in other words, it has caused energy to be expended. We shall speak of the source of this energy in a moment. The expression for the work done is simply the pressure into the increase in volume, namely, *pdv*. We see that this term actually represents work, for it has the correct dimensions, ML^2 namely, -T-, and may be ultimately expressed as force x distance.

The source of the energy which has been used in the work of expansion if the temperature has been maintained constant, is the heat which has been drawn from the surroundings. This is an " isothermal expansion ". If we had enclosed the gas in a " heat tight" case and allowed it to do work we would have found that its temperature would have fallen (though only to an infinitely small extent, if the expansion was infinitely small). In the latter case the work is done at the expense of the heat energy of the gas itself.

By this heat energy we mean the kinetic energy of the molecules, for heat, properly speaking, does not reside *in* a system, being transferable from one system to another, but always undergoing transformation into something else *(i.e.* say, into kinetic energy of the molecules) after its addition to a system.

It will be observed that in order to keep the temperature of the gas constant whilst expansion proceeds we have assumed that heat is given to the gas from its surroundings in order to compensate for the cooling effect of the expansion itself. This really involves the principle of the First Law, for what we have actually done is to transform heat energy into mechanical energy, the gas simply acting as the medium.

Besides heat and mechanical energy we are acquainted with many other forms of energy, *i.e.* electrical energy, magnetic energy, radiant energy, surface energy (*i.e.* the product of surface tension of a liquid into the area of the surface), and chemical energy. Now, the First Law deals with the transformation of any kind of energy into any other kind, and according to the law, when a certain amount of one kind disappears an exactly equivalent quantity of some other kind must appear. When a bullet strikes a target there is heat produced, the amount of heat being equivalent to the kinetic energy which the bullet possessed just before striking.

If we lose 100 ergs of mechanical energy in some process we must gain 100 ergs in some other form – heat, for example; for energy can neither be created nor destroyed as a whole. This is why perpetual motion is impossible; it is more accurate to say that in the experience of mankind, no instance of perpetual motion has been recorded, and we have therefore concluded that such is impossible. This is stated by the First Law. The First Law is a law of experience, just as the Second Law is, as we shall see later.

Assuming the validity of the First Law, we are led to make a determination of the so-called "mechanical equivalent of heat," an experiment carried out in the first instance by Joule, and later by Rowlands and others. Joule's determination of the equivalent consisted in placing some water in an isolated vessel and then rapidly rotating a paddle in the water. He observed that the temperature of the water rose – just as Rumford, many years previously, had observed that in the boring of a cannon the friction caused the metal to become extremely hot. With Joule's arrangement it was possible to obtain a quantitative connection between the mechanical work done in stirring, and the heat added to the water, as evidenced by the rise in temperature. The paddle was worked by means of a falling weight, the working agency being therefore gravitation. Frictional effects were reduced as far as possible.

Knowing the weight of the falling body, and the distance through which it fell, the amount of work done by gravitation can be directly obtained. If, now, the rise in temperature of the water is measured, the heat capacity of the water being known, the amount of heat in calories generated by the motion of the paddle is also obtained. If we neglect or allow for friction and other disturbing effects we can find how many mechanical work units, *i.e.* ergs, are equivalent to i heat unit, *i.e.* i calorie. Joule found that i calorie — 4.2×10^7 ergs (approx.). Rowlands's more accurate determinations agreed closely with Joule's value. In the cases where work is done by the expansion of a system (against pressure), we have assumed that the temperature *can* be kept constant by adding the requisite amount of heat which is transformed into mechanical work *(pdv)* via the system considered.

That is to say, we assume the validity of the First Law regarding the mutual transformation of heat into mechanical energy, and *vice versd.* To give precision to our statement of the First Law we want to be able to express it algebraically. For this purpose let us think of some system (solid, liquid, gaseous, homogeneous or heterogeneous) which undergoes some physical or chemical change. In general there will be associated with this *matter* change the following *energy* changes:

1. A certain amount of heat may be absorbed or evolved by the system.
2. A certain amount of external work is either performed *by* the system or upon it (by the surroundings).
3. The internal energy of the system may have increased or de- creased. (This naturally possesses a complicated mechanism. Often it can be regarded as due to changes in the relative position, rate, and kind of motion of the molecules. We are, however, not concerned here with the mechanism of changes going on in the interior of a system. We simply consider the fact itself that the internal energy, or, as it is often called, the total energy, has altered in a given process.)

No matter what the process itself may be, the First Law tells us that decrease in internal energy (supposing there to be a decrease, which *decrease* we denote by + U) must be equal to the external work done, namely, A (say due to an increase in volume against a pressure) plus the amount of heat evolved or lost (call the heat absorption + Q, then - Q represents the heat evolved).

Consider a system in which a chemical change occurs without alteration in volume, and without doing any form of external work. Suppose the internal energy decreases by the amount U. This energy leaves the system in the form of heat evolved, which is denoted by the term – Qv the suffix *v* indicating that the volume has been kept constant.

Even when the reaction cannot be carried out without a change in volume (say an increase in volume) the heat equivalent of the work done by the system in expanding, against the atmosphere, say, can be calculated, and this amount is added to the observed heat evolved to give the quantity - Q The symbol Q,, must not be confused with Q. As a matter of fact, Q is in many cases (though not always) a much smaller quantity than Q_w. Q is often called a " latent heat".

If a process occurs and does work A of any kind (not necessarily mechanical), and at the same time the internal energy of the system diminishes by the amount U, then the term Q simply stands for the *difference* of the two terms A and U. Q is not a measure of U at all. In some cases (A - U) may be almost zero, *i.e.* A and U may be nearly equal, and therefore Q may be nearly zero, and this may be so when both A and U (or $-Q_v$) possess large numerical values. In other cases, however, U (or - Qt,) may be very small, in which case the work done (A), which may be great, is at the expense of the heat taken in, *viz.* Q.

This is very nearly realized in the work done by an expanding gas. Naturally these terms must be expressed in the same units, say calories. Note also that in the above nomenclature + U represents a *decrease* in the internal energy and - U therefore represents an *increase* in the internal energy of the system. (The sign here has significance with respect, not to the system itself, but to the surroundings. A gain in energy to the surroundings must mean loss in energy to the system.) The term + A represents external work done *by* the system on the surroundings, — A represents external work done *upon* the system by the surroundings.

+Q represents heat added to or absorbed by the system.

—Q represents heat taken from or evolved by the system.

THE CONCEPT OF MAXIMUM WORK

In returning to the question of the work done in the expansion of a gas, it will be observed that we considered an increase in volume so small that the pressure of the gas had remained constant. Suppose now that we consider a finite change in volume from v to 0_2. Since the system is a gas we know that on increasing its volume its own pressure will decrease. The work which can be got out of the gas will depend upon the opposing pressure, and it is clear that it depends upon the magnitude of this opposing pressure whether the gas can expand up to v_2 or not. If the opposing pressure is always less than that possessed by the gas, even at the large volume »$_{21}$ the piston on being released will move rapidly from *v±* to *v%*, the work done being*p(v% - Vi)*. But this work is not in any sense a definite amount characteristic of the pressure of the gas, for/ is the external opposing pressure. It is evident that the work done by the gas can vary *(even when expanding between the same volume limits depending on the values of the opposing pressures*. If the opposing pressure is zero the work will be zero.

Is there an upper limit to this work term? To get at this we have to consider the work done from the standpoint of the pressure possessed by the gas itself. Let us first of all ask the question, What are the conditions under which a gas must be placed in order that a very small expansion *dv* may be accompanied by a *maximum* output of work ? *The necessary condition which must be fulfilled so that the gas can do a maximum amount of work in expanding by dv is that the external pressure should be less than the pressure (call this now p) possessed by the gas by an infinitely small amount dp.*

If we imagine the outer pressure (that of the atmosphere) to be *(p - dp* then the pressure of the gas/ will *be just* able to overcome this outer pressure. The work done is *(p - dp) dv,* or, what is the same thing, *pdv,* for the product *dp. dv* is only of the second order of magnitude, being the product of two infinitely small quantities. If the external pressure had been exactly equivalent to *p* the piston would have remained motionless. If the external pressure had been / + *dp* the piston would have moved inwards infinitely slowly, the work done *upon* the gas being - *(f + dp)dv* or - *pdv,* the negative sign denoting a decrease in the volume of the gas. It will be seen that under the above pressure conditions the process is a *reversible* one in the sense that for the same volume change (in one case an increase, in the other a decrease) the work done is represented by the same *numerical* magnitude + *pdv.* When the expansion takes place in this way it is accompanied by maximum work, so that the production of maximum work is the characteristic of a process which is being carried out in a *reversible* manner.

We shall speak at greater length of reversible processes later on. It may here be noted that since the external pressure is so arranged that the pressure possessed by the gas is just able to overcome it, the expansion through *dv* must take place infinitely slowly. Hence the gas does not get up momentum which on stopping would be transformed into heat of indeterminate amount. No friction effects take place either on the outgoing or return journey. The process can, therefore, be repeated an infinite number of times without *way permanent* drain on the energy resources of the system and surroundings. As already mentioned the source of the energy which enables the gas to expand and do work and yet keep the temperature constant is then the " heat content" of the surroundings. In a reversible process the heat taken in from the surroundings during the volume expansion is *exactly* the same in amount as the heat given back to the surroundings when the gas contracts isothermally and infinitely slowly to its original position.

If there had been a net loss or gain of heat, say, as a result of the total operation, the process is called an irreversible one, for on completing the process isothermally there would be a permanent change in the " heat content" of the system and surroundings. A reversible process is essentially one in which the change takes place in a known and definite manner, there being no " accidental'* energy transformations taking place, dependent on *the path* followed, *i.e.* dependent on the physical structure, shape, or properties of the system. Of course a reversible process is only a limiting case. It cannot be realized in practice since a "frictionless piston," an "infinitely slow process" with a "pressure difference infinitely small" on the sides of the piston, is only *mentally* realisable.

The significance of reversible processes and maximum work production will be clear when we come later to study the Second Law of Thermodynamics. We have been considering how the maximum work can be done by a gas expanding through an infinitely small volume *dv.* We have now to consider the maximum work produced when a gas undergoes a measurable volume increase from. We know that the pressure of a gas falls as the volume increases. We cannot, therefore, use the expression as the expression for the maximum work. In fact, we cahnot think of the external pressure on the piston remaining constant. It must also continuously decrease in the same.proportion as that of the gas pressure itself, being at any stage only less than that of the gas itself by the infinitely small amount *dp.* To find out what the maximum work is we have to suppose the total expansion carried out in a series of infinitely small steps, each being represented by the product of the existing pressure into a small volume increase *dv* and then add them all together.

It may be pointed out, parenthetically, that in the isothermal evaporation of a liquid we have a case in which there may be a large volume change, the pressure remaining constant, namely, the pressure of the saturated vapour.

Suppose a cylinder fitted with a piston contains some liquid and saturated vapour. Let us consider what is the value of the maximum work which will be done when one gram-mole of the liquid is vaporised. We suppose that there is more than one gram-mole of liquid present to start with, so that throughout the whole operation the vapour is saturated.

Suppose the piston is raised in a reversible manner, *i.e.* infinitely slowly, the pressure outside being very nearly the value of /, the pressure of the saturated vapour (the pressure outside the piston) is - *dp*. The process is assumed to take place isothermally, and hence heat must be continuously supplied from the outside to supply that which has become latent in the process of vaporisation. If the increase in volume of the vapour corresponding to the production of one gram-mole of vapour is V, and the decrease in the volume of liquid at the same time is v, the maximum external work done by the system as a whole is /(V - v). In general we can neglect v compared to V.

THE SECOND LAW OF THERMODYNAMICS

The Second Law is a relationship which has meaning and significance (in the first place) for a *cycle* or a cyclic process. Consequently it is with such processes that we are immediately concerned.

It will be noticed that in the definition of the First Law given above, stress has been laid on the words " *if* or *whenever*" heat is converted, etc. Experience has shown that work is always capable of being transformed into heat, but this is not always the case for the opposite transformation, that of heat into work in a cyclic process. Thus whilst the law of the Conservation of Energy furnishes us with the relationships which must necessarily hold when one form of energy is transformed into another, the so-called Second Law brings out a peculiarity which belongs alone to energy in the form of heat as regards its convertibility in a cyclic process into other forms of energy.

The Second Law as a statement of the results of experience teaches us that in a completed cycle there are certain definite limitations in connection with the transformation of heat into other forms of energy. Qualitatively one may state it thus:— External work, such as the kinetic energy of moving bodies, may be transformed in many ways, and *completely* into another form, heat for example; but the conversion of heat into work by an engine working in a cycle is not a complete one, and may not even be possible to the slightest extent. Remember this does not in any way contradict the First Law. For whether the heat transformed be small or great die First Law holds absolutely, in that this small or great heat change is transformed quantitatively into some equivalent form of energy, say motion.

The First Law simply states that if or whenever heat is transformed an equivalent of some other energy appears. It does not state that heat under all conditions *can* be transformed. The distinction is most important. As an example of heat that no one has been able to utilise, transform, we might take the heat energy of the ocean.

We know that the ocean is at a certain temperature, and that it requires energy to be added to it to raise its temperature. Hence it must contain energy in the bound state. No one has by any cyclic device been able to use the heat energy, say to drive the screw of a vessel in the ocean itself. There are numerous other instances, and we have gradually come to the conclusion that for some reason or other an engine (working, as an engine must work, namely, in a cycle) cannot *always* convert heat into work. This naturally leads to the question: What is the condition which determines this conversion or non-conversion of heat into work in a cyclic process? The answer is simple, it is a question of temperature or rather temperature differences. We cannot convert heat into work unless we allow this heat to pass from a high temperature to a lower one. The efficiency of the cyclic process, the fraction of heat converted into work, is dependent on the temperature difference between the hottest and coldest part of the system. This is why the heat of the ocean is unavailable; we ourselves are living at the same temperature (or even higher).

The ship, for example, is at the same temperature as the ocean, and hence, if the above principle is correct, one cannot expect the heat of the ocean to be utilisable by the ship. On the other hand, the inhabitants of a colder planet might readily utilise our ocean heat, and, indeed, the heat of our planet itself, for work, if only the mechanical difficulties could be overcome. In asserting that heat cannot in general be converted completely into work, it is absolutely essential to realize that we are dealing with a cyclic process, *i.e.* the kind of process, or rather series of processes which any thermodynamic engine is capable of carrying out. The limitation in respect of conversion would be meaningless unless we are thinking of a completed cycle. For such a case, when we speak of work, we mean not the work done in any one of the various stages passed through, but the net amount of work obtainable from the completed cycle it is only with such net work that the Second Law deals.

To show the necessity of keeping to the idea of a completed cycle when discussing the Second Law, let us consider not a cycle, but a single stage by itself. Take the case of a perfect gas expanding iso-thermaliy, and doing work, the maximum value of which is RT log Vg/Vj per mole at the absolute temperature T. We have already seen that, in order that this process may occur isothermally, it is necessary to draw upon the heat content of the surroundings, the surroundings being themselves at temperature T. That is, heat is being converted into work at constant temperature. Further, since the gas is a perfect one, we have already seen that the heat taken in to maintain the temperature constant is *completely* converted into work. Hence to such a process (which may well form a *step* in a cycle) the Second Law cannot be applied since the process is not a completed cyclic one, in which the system returns to the same physical and chemical state as characterised it at the beginning.

The Second Law, as stated by Clausius, is as follows:—"It is impossible for a self-acting machine working in a cycle, unaided by any external agency, to convey heat from a body at a low temperature to one at a higher temperature; or heat cannot *of itself* (*i.e.* without the performance of work by some external agency) pass from a cold to a less cold body."

First note the phrase "of itself" — heat cannot of itself pass from a low to a high temperature. We have seen that by means of an adi-abatic compression of a system the temperature rises, heat being evidently thereby carried up the temperature scale. But this does not contradict the above statement about the *natural* flow of heat, for we have, by the aid of external agency, had to do work, namely, to compress the system in order to make the temperature rise. It is a known experimental fact that systems, say gases, naturally tend to expand, the molecules tend to fly apart and not to contract. This fact is evidence, of a molecular kind, for the Second Law, for by expanding the system cools. So much for the qualitative statement of experience.

We now come to the question of a quantitative statement. It must be clearly understood that the quantitative formulation of the Second Law (which will be given in a moment) holds *only for a reversible cycle* of changes. We have already discussed changes of this nature. One other instance may be given. Suppose a system of any kind — an engine, as it is called — takes in heat Q_1 from the boiler at temperature *TI* and drives a piston and crank, thereby doing external work, and then gives out heat Q_2 to the condenser at temperature T_2, the amount of heat given out to perform work is $Q_1 - Q_2$ and this will be quantitatively the work done; for friction is entirely absent, as we suppose it to be an ideal engine. If, now, some external agency performs a quantity of work on this engine numerically equivalent to Q_x - Qg, and if the engine thereby takes in heat Q_1 at T_2 and gives up Q_1 at the high temperature T then the engine is a reversible one. When the engine is working directly it must be doing maximum work, just able to overcome opposing forces in order to comply with the condition of working reversibly.

Chapter 14

Liquification of Gases and Critical Phenomena

In the gaseous state the molecules of matter are relatively far apart and in perpetual rapid motion, and this motion is sufficient to overcome the attractive forces which exist between the molecules. These, therefore, lead an existence of practically complete independence. As the temperature of a gas is lowered, the kinetic energy of the molecules may be reduced to such an extent that the more sluggish molecules (for all do not move with the same velocity) are no longer able to overcome the force of mutual attraction or cohesion. They cohere to one another, therefore, and a liquid is formed. This process of condensation or liquefaction will be helped by an increase of pressure (or diminution of volume of the gas) as well as by reduction of the temperature.

The conditions of temperature and pressure necessary to bring about liquefaction vary greatly for different gases, for while sulphur dioxide can be liquefied, under atmospheric pressure, at a temperature of — 8°, carbon dioxide must be cooled to below —78-2°, and oxygen to below -182*5°. Since however, pressure assists condensation, sulphur dioxide can be liquefied even at 20° if the pressure is increased to 3-24 atmospheres.

Liquefaction of Gases.—While the attempts of the earliest workers—of whom the most important was Michael Faraday (1791-1867), of the Royal Institution, London—to liquefy gases were crowned with a large measure of success, a certain number of gases, *e.g.*, hydrogen, oxygen, nitrogen, methane, proved refractory. All attempts to obtain these substances in the liquid state having proved unsuccessful, even when the pressure was raised to 2790 atmospheres, the gases were regarded as being non-liquefiable and were therefore spoken of as *permanent gases.* In 1869, however, the reason for the failure to liquefy these gases became clear when, as will be discussed more fully presently, Thomas Andrews, (1813-85) Professor of Chemistry in Belfast, showed that for each gas there exists a temperature, the *critical temperature,* above which the gas cannot be liquefied no matter how great may be the pressure.

After the conditions necessary for the liquefaction of gases had been established, Raoul Pierre Pictet (1842-1929), in Geneva, and Louis Paul Cailletet (1832-1913), Ironmaster at Chatillon-sur-Seine, succeeded in liquefying a number of the so-called permanent gases, the former by cooling the gas, under pressure, by means of liquid carbon dioxide boiling under reduced pressure, and the latter by rapidly reducing the pressure and so allowing the compressed gas to expand adiabatically.

In the latter case the gas, on expanding against the pressure, does a certain amount of work, and the heat corresponding to the work is taken from the gas itself. The temperature of the gas, therefore, falls, and may reach the point of liquefaction. This principle of adiabatic expansion is employed in the Claude process for the liquefaction of air on an industrial scale.

When a gas is compressed adiabatically, its temperature will be raised. In 1853 it was shown by Sir William Thomson (Lord Kelvin) (1824-1907), Professor of Natural Philosophy in the University of Glasgow, and James Prescott Joule (1818-89), an English brewer, that when compressed air, and certain other gases, at temperatures between 0° and 100°, are allowed to expand through a porous plug or valve, the temperature falls; but that in the case of hydrogen a rise of temperature takes place.

This heat effect on expansion is known as the *Joule-Thomson effect,* and it was later shown that there is, for each gas, a certain temperature at which the sign of the heat effect undergoes change. This temperature is known as the *temperature of inversion of the Joule-Thomson effect.* When, therefore, a compressed gas is allowed to expand at a temperature below the temperature of inversion of the Joule-Thomson effect, cooling takes place; and the degree of cooling is all the greater the lower the initial temperature of the compressed gas. For air (under a pressure of 200 atmospheres) the temperature of inversion of the Joule-Thomson effect is 240°; for hydrogen it is - 79°, and for helium, - 173°.

In 1895, C. von Linde, in Germany, and W. Hampson, in England, patented apparatus in which the Joule-Thomson effect is applied to the liquefaction of air and other gases. In these apparatus the compressed gas is allowed to expand at a nozzle and the cooled gas is caused to pass around the tube by which the compressed gas reaches the nozzle (Fig. 16). The incoming gas is thus progressively cooled, and at length a temperature is reached such that when the gas expands the ensuing fall of temperature is so great that the gas is liquefied. By this method even hydrogen and helium have been obtained in the liquid state.

The Joule-Thomson effect is due to the fact that gases do not obey Boyle's law and Joule's law. For moderate increase of pressure at the ordinary temperature, most gases, we have seen, are more compressible than according to Boyle's law; that is, the value of the product #w at the higher pressure is less than its value at the lower pressure, owing to the attraction between the molecules. When, therefore, expansion takes place, work has to be done against the attraction of the molecules, and the heat corresponding to this work is taken from the gas itself: its temperature, therefore, falls. In the case of hydrogen at the ordinary temperature, however, the product *pv* has a greater value at a higher pressure than it has at a lower. When, therefore, the compressed gas is allowed to expand there is a decrease in the value of *pv,* as if work were done on the gas, and this work shows itself in a rise of temperature.

The Joule-Thomson effect also is due to the fact that real gases do not obey Joule's law which states that the total internal energy of a given mass of a perfect gas at constant temperature is independent of the volume. In an actual gas, on expanding internal work is done in overcoming the cohesion between the molecules.

Critical Phenomena.—" On partially liquefying carbonic acid by pressure alone, and gradually raising at the same time the temperature to 88° Fahr., the surface of demarcation between the liquid and gas became fainter, lost its curvature, and at last disappeared. The space was then occupied by a homogeneous fluid, which exhibited, when the pressure was suddenly diminished, or the temperature slightly lowered, a peculiar appearance of moving or flickering striae throughout its entire mass. At temperatures above 88° no apparent liquefaction of carbonic acid, or separation into two distinct forms of matter, could be effected, even when a pressure of 300 or 400 atmospheres was applied. Nitrous oxide gave analogous results." So wrote Thomas Andrews in 1863.

The occurrence of a critical temperature at which liquid and vapour become identical was established by Andrews in the course of his study of the pressure-volume relations of carbon dioxide at different temperatures, The behaviour will best be understood with the help of the diagram, in which pressure is plotted against volume.

When one considers the isothermal curve for 13-1°, it is seen that as the pressure increases the volume of the gas diminishes, until, at a certain pressure (49-8 atmospheres), liquefaction takes place (point B). On diminishing still further the volume of the gas (vapour), the pressure remains constant, until the whole of the vapour has been condensed to liquid (point C). Thereafter, the liquid being only slightly compressible, a great increase of pressure is required to effect even a slight diminution of volume. The pressure-volume curve, therefore, rises almost vertically. At higher temperatures, *e.g.*, 21 -1°, a similar behaviour is found, but as the temperature is raised, the pressure at which liquefaction takes place becomes higher and the length of the horizontal portion of the isotherm becomes shorter and shorter until, at a certain temperature, it dwindles to a point. The isothermal curve then becomes continuous. This is shown by the isotherm for 31-1°. At this temperature, although a flattening of the curve is shown, there is no horizontal portion and no abrupt change from the gaseous to the liquid state.

At temperatures above 31-1° the flattening of the curve becomes less marked and, at 48°, is no longer observable. Andrews was thus able to show that for carbon dioxide there is a certain temperature, which he called the *critical temperature*, below which the gas can be liquefied by increase of pressure, but above which liquefaction cannot take place no matter how great the pressure may be. Other behave in a similar manner.

At the critical temperature the vapour pressure of the liquid must have a definite value, and this is known as the *critical pressure;* at this temperature, also, the substance must have a definite density *(critical density)*, the reciprocal of which is called the *critical volume.*

DETERMINATION OF THE CRITICAL CONSTANTS

The critical temperature of a substance which is liquid under ordinary conditions can be most easily determined by sealing up a quantity of the liquid in a strong glass tube from which all air must be removed. The tube is then heated in a suitable oven, the temperature of which can be slowly raised. As the temperature of the critical point is approached, the liquid meniscus becomes billowy and indistinct, and then suddenly disappears altogether when the critical temperature is reached.

On cooling through the critical temperature, formation of a mist and a liquid meniscus is observed. In the case of a gas the substance may be compressed in a capillary glass tube until liquid is formed. The tube is then heated until the meniscus disappears, the pressure being increased during the process so as to keep the meniscus in a fixed position.

For the determination of the critical pressure, the method due to Michael Altschul may be employed. The substance (liquid) is contained in a strong glass tube *d*, the lower end of which is closed by a movable thread of mercury *e*, and the tube is connected with a pressure pump.

The upper end of the tube is heated to above the critical temperature, the pressure being at the same time increased so as to maintain the liquid in the tube. The lower end of the tube is kept at the ordinary temperature (that is, below the critical tem- perature), and, therefore, at some intermediate point the temperature must be the critical temperature. At that point a meniscus will be formed, and the critical pressure will be produced. This pressure can be read off on the manometer attached to the pump.

The critical volume is usually determined indirectly by making use of the *law of rectilineal diameters* discovered by Cailletet and Emile Mathias, Professor of Physics in the University of Clermont, France. If in a system of rectangular co-ordinates the density of the liquid and of the vapour is plotted against the temperature, a curve is obtained.

This curve shows clearly that as the temperature rises the density of the saturated vapour increases and the density of the liquid decreases, the density of liquid and vapour becoming identical at the critical point. If, also, one plots the means of the densities of the liquid and vapour, Cailletet and Mathias found that these values lie on a straight line CD, which passes through the critical point. This is the *law of rectilineal diameters.*

In order to determine the critical density or critical volume of a substance, therefore, one determines the density of the liquid and of the saturated vapour at a series of temperatures, *not too far removed from the critical point,* and one plots these values against temperature in rectangular co-ordinates. The corresponding values of density are joined by straight lines; these lines are bisected, and, through the points of bisection, a line is drawn until it cuts the perpendicular *ab,* drawn from the critical temperature. The point of intersection *c* gives the value of the critical density, and the reciprocal of this is the critical volume.

CONTINUITY OF THE LIQUID AND GASEOUS STATES

Although one is accustomed to think of the liquid and gaseous states as being sharply distinguished one from the other, a consideration of the critical phenomena which have just been discussed will show convincingly that one is scarcely justified in so thinking. A liquid and its vapour when heated together approach *continuously* to each other in their properties and become identical in all respects at the critical point: there is no abrupt change of properties. Moreover, one can pass continuously from the liquid to the gaseous, and from the gaseous to the liquid state without there being at any time a discontinuity in the properties of a substance, and without there being any separation into two phases or distinct forms of matter.

When the ends of the horizontal portions of the isotherms are joined together, a so-called " border curve " is obtained. At the summit of this curve lies the critical point K, and within this curve lie all the values of pressure, volume, and temperature at which liquid and vapour can coexist; outside this curve only homogeneous systems, liquid or vapour alone, can exist. Suppose, then, that carbon dioxide gas, under atmospheric pressure, is heated to a temperature above the critical temperature, say to 48°C, and that it is then subjected to a pressure greater than the critical pressure, say, to 100 atmospheres, the system will change in volume, as shown by the isotherm for 48°C.

If the temperature be now lowered to 13-1°C, while the pressure is maintained at 100 atmospheres, a point on the curve CD will be reached, a curve which applies to *liquid* carbon dioxide. At no point in the process is there any abrupt change of state; the change from gas to liquid is continuous. Similarly, one may start with liquid carbon dioxide under the conditions represented, say, by the point C. If the pressure on the liquid is increased, say, to 100 atmospheres and the temperature raised to 48°C, the condition of the system will be represented by a point on that isotherm; and on reducing the pressure to that of the atmosphere and then allowing the temperature to fall to 13-1°C, we shall have the carbon dioxide as a gas, and at no time during the whole process will any discontinuity of properties be observed, or any separation into liquid and vapour. The passage from the liquid to the gaseous state is continuous.

LIQUIDS, AND THEIR PROPERTIES

VAPORISATION AND VAPOUR PRESSURE

Liquids differ from gases in that they do not fill completely the whole space which may be offered to them; they possess no definite form, but, owing to cohesion, they have a definite volume. The compressibility of liquids, which is small, and their expansion with temperature depend on the nature of the liquid. The volume of liquid is, therefore, not a colligative property.

Although, in the case of liquids, no such simple relations obtain as in the case of gases, the subject of the molecular weights of substances in the liquid state will be briefly reviewed, after the physical properties on which much of our knowledge of the molecular weights of liquids chiefly depends, have been studied.

STRUCTURE OF LIQUIDS

Until a few years ago the liquid state was considered to be similar to the gaseous state, although, owing to closer packing of the molecules, the intermolecular forces are necessarily greater than for gases. As in the case of gases, so in the case of liquids, the molecules were regarded as being distributed in a disordered or random manner. This view, however, which might have persisted if one could have regarded the molecules as being completely symmetrical spheres, has had to be abandoned.

The introduction of the conception of dipoles and the conclusion, based on experiment, that molecules may have a geometrical shape—linear, like carbon dioxide, or flat, like benzene— has necessarily led to the belief that such molecules must tend to pack in some sort of ordered array. Notable examples of the orderly parallel wise arrangement of molecules in the liquid state are found in the case of the turbid liquids formed by a number of compounds with long-chain molecules (*e.g.*, azoxyanisole). Although X-ray examination has shown that no space-lattice in the ordinary sense of the term exists, some kind of asymmetry must be present as these liquids are anisotropic and show the property of double refraction

Although " ideal liquids," or assemblages of closely packed spheres, oscillating over a certain range, do exist (in the case of fused metals, for example), evidence of the existence of a structure in some liquids, at least, apart from those just referred to, is given by the diffraction of X-rays, the study of which has so greatly advanced our knowledge of the crystalline state. It has been shown, in fact, that there is a general similarity between the X-ray patterns for liquids and those of the corresponding solids; and so there has gradually grown up the view that there is a tendency for a liquid to imitate the solid in general atomic or molecular arrangement.

The diffraction of X-rays has shown that water, for example, has a marked structure, based on a tetrahedral arrangement of the molecules. In ice, the tetrahedra are arranged in a fairy open structure, like the tetrahedra of tridymite, one of the crystalline forms of silica. When ice melts, however, there is a partial changeover to the more compact arrangement shown by quartz, and the relative amount of the " quartz " structure increases to a maximum as the temperature is raised to 4°. As the temperature is raised above 4°, there is a progressive breakdown of the " crystalline " structure with an increasing randomness in the distribution of the molecules. Above 150° the random distribution is complete and the molecular arrangement approximates to that of an " ideal " liquid.

It seems clear, therefore, that in some liquids, at least, the molecules are arranged with some degree of orderliness; and such packing of the molecules in ordered array, and the formation of polymerised or associated molecules have an influence, as we shall learn, on the physical properties of liquids, such as viscosity, heat of vaporisation, etc.

VAPORISATION AND VAPOUR PRESSURE

Although the molecules of a liquid, as a whole, are unable to overcome the force of cohesion, yet the molecules are not all moving with the same velocity, and it may happen that some, having a higher velocity than the average, are able to tear themselves away from the general mass and escape into space, where they move about freely as the molecules of a gas or vapour. The liquid evaporates.

If the liquid is placed in a closed vessel so that the escaping molecules cannot pass away into unlimited space, it is found that a pressure is produced, as can readily be shown by breaking a bulb of ether in a closed bottle fitted with a manometer or pressure gauge. The pressure thus produced, however, does not go on increasing indefinitely. As more and more vapour is produced, some of the molecules will strike the surface of the liquid and be recaptured; and the number of such molecules will increase as the pressure or concentration of the vapour molecules increases.

As more and more vapour is formed, therefore, a point will at length be reached when the number of molecules recaptured by the liquid will be equal to the number which escape from the liquid. A state of equilibrium will thereby be attained, and the pressure of the vapour in this state of equilibrium is known as the *vapour pressure* of the liquid. The vapour pressure, or pressure of the *saturated vapour* as it is sometimes called, increases with the temperature, but has, at each temperature, a definite value.

Although, at any given temperature, the vapour pressure of different liquids may differ greatly—some liquids being, as it is said, more volatile than others—yet, when one plots the values of the vapour pressure against the temperature, it is found that the curve so obtained *has the same form for all liquids,* and is convex to the temperature axis.

When the temperature is raised the vapour pressure increases, and when the pressure of the vapour becomes equal to the atmospheric pressure or to the external pressure on the liquid, bubbles of vapour are formed within the liquid and the liquid is then said to boil. The *boiling-point* of a liquid is therefore the temperature at which the vapour pressure of the liquid becomes equal to the external pressure on the liquid. If the external pressure is increased, the boiling-point is raised; and if the external pressure is diminished, the boiling-point is lowered. A liquid will boil at a lower temperature on the top of a mountain than it will on the seashore.

For this reason, substances which undergo decomposition at a temperature below the boiling-point under atmospheric pressure can frequently be distilled at a lower temperature, without decomposition, by carrying out the distillation under reduced pressure. This fact is widely taken advantage of for the purification of substances both in industry and in the scientific laboratory.

DETERMINATION O! THE VAPOUR PRESSURE OF A LIQUID

For the determination of the vapour pressure various methods, both static and dynamic, may be employed.

Dynamic Methods

Among the first to determine the vapour pressure of liquids with accuracy by a dynamic method were William Ramsay and Sydney Young (1857-1937). According to this method, one maintains a constant pressure and ascertains the temperature at which the liquid boils under that pressure.

The bulb of the thermometer, which passes through a rubber bung, is covered with cotton-wool, and the lower end of the dropping funnel is drawn out to a point and bent so that it just touches the cotton-wool. The boiling tube is supported in a suitable heating bath . There is a tube to condense vapour, is a large bottle which acts as a reservoir and serves to maintain a steady pressure, and a manometer.

A quantity of the pure liquid is placed in and the pressure within the apparatus is reduced by means of an exhaust pump to a definite value, say 30 mm. of mercury. Liquid is allowed to flow slowly from the funnel so as thoroughly to moisten the cotton-wool surrounding the bulb of the thermometer. As evaporation takes place, the temperature recorded by the thermometer will alter and will reach a steady value, if the pressure in the apparatus is kept constant. This temperature is the boiling-point of the liquid under the pressure within the apparatus; or, in other words, the pressure shown by the manometer represents the vapour pressure of the liquid at the temperature recorded by the thermometer.

The pressure within the apparatus is then altered by allowing air to enter through the tap, and the boiling-point under the new pressure determined. In this way the vapour pressure at a series of different temperatures can readily be determined. The boiling-point method introduced by Alexander Smith (1865-1922), Professor of Chemistry, Columbia University, and Alan W. C. Menzies, of Princeton University, is somewhat simpler than that of Ramsay and Young, and requires the use of only a small quantity of liquid.

The liquid under investigation is placed in a small bulb-tube A, which is then attached by thread to the thermometer. Tube is nearly filled with a liquid of considerably higher boiling-point than that under investigation. Tube is connected by means of with a manometer and pressure-regulating bottle. The temperature of the liquid in second tube is regulated by means of the bath.

To carry out a determination, the second tube is heated to a constant temperature, say 15°C, and the pressure in the apparatus is then reduced. As the pressure falls, air escapes from the bulb-tube. Later, the liquid in A begins to boil, and a stream of bubbles escapes from the capillary. Air is now allowed to enter slowly through a very fine capillary,until the liquid in another tube just ceases to boil. At this point the temperature on the thermometer and the pressure are read. Similar determinations are then carried out at higher temperatures.

Static Method

The vapour pressure may also be determined by means of the *isoteniscope.* This instrument, we owe to Smith and Menzies. The liquid under investigation is introduced into the isoteniscope in such amount that the bulb is about half full, and the liquid stands about half-way up the limbs of the U-tube. The apparatus is placed in a suitable heating bath and is connected with a manometer and pressure-regulating bottle. When the temperature has been adjusted to a definite value, the pressure is reduced until the liquid in the bulb of the isoteniscope begins to boil. Air is then allowed to enter very slowly through the tap H, until the level of liquid is the same in the two limbs of the U-tube. The pressure registered by the manometer is the vapour pressure of the liquid at the temperature of the bath.

LATENT HEAT OF VAPORISATION

When a liquid is boiling freely, its temperature remains constant no matter how much heat is added to the liquid; and since the added heat brings about no rise of temperature, it is said to become *latent.* The amount of heat required to convert 1 g. of substance from liquid to vapour, without producing an alteration of temperature, is called the *latent heat of vaporisation.* This latent heat is the equivalent of the work done (1) in increasing the volume against the constant external pressure, and (2) in overcoming the mutual attraction of the molecules. In the case of associated liquids, the latent heat of vaporisation will also include the heat of dissociation of the complex molecules, and will therefore have a higher value than in the case of non-associated liquids.

Conversely, when vapour condenses to liquid, an amount of heat equal to the heat of vaporisation is given out.

When a liquid evaporates spontaneously, the heat of vaporisation is taken, in the first instance, from the liquid itself, and the temperature therefore falls. This effect is readily observed if one sprays ether, or other very volatile liquid, on the skin. Through the rapid evaporation of the liquid the temperature may fall to such a degree that local anaesthesia is produced.

Similarly, by the rapid vaporation of water the temperature may be lowered to such an extent that the water freezes. (This is readily demonstrated by causing water in a Dewar vacuum tube to boil rapidly under reduced pressure.) From very early times, in fact, ice has been obtained by the natives of India by exposing water in shallow pans, embedded in straw to prevent access of heat. On a clear night, evaporation takes place so rapidly that the water freezes. The lowering of temperature produced by the evaporation of volatile liquids, *e.g.*, ammonia, sulphur dioxide, is made use of in various types of refrigerators.

This cooling by evaporation is readily explicable on the basis of the kinetic theory, for, during evaporation, the molecules which overcome the mutual attraction and pass into free gas molecules are those which possess a relatively large amount of energy. The average kinetic energy of the remaining molecules, therefore, will be decreased, and the temperature, which is measured by the kinetic energy, will fall.

The latent heat of evaporation of water (or the heat of condensation of steam with reversed sign), which is of so great importance in connection with the use of the steam engine and with the transport of heat by means of steam, was first determined in 1764 by Joseph Black (1728-99), Lecturer in Chemistry in the University of Glasgow, who passed steam at a known temperature into water contained in a calorimeter. From the rise of temperature produced in a known weight of water by a known amount of steam the latent heat of vaporisation could be calculated. For water, the latent heat of vaporisation at 100° is 539-55 calories per gram.

VISCOSITY

One of the properties characteristic of liquids is that of flowing; but it is a well-known fact that some liquids flow much more readily, or are much more mobile, than others.

This can be demonstrated by sealing up ether, water, and glycerol in separate glass tubes, 35 to 40 cm. long and about 7 mm. in diameter. The tubes should be about four-fifths full of liquid. If all three tubes are inverted at the same time, it is found that the ether flows down the tube more rapidly than the water, and that, at the ordinary temperature, glycerol is so viscous that it flows with very great slowness. That the viscosity diminishes with rise of temperature is readily shown by heating the tube of glycerol in a bath of boiling water. The glycerol is now quite mobile.

When water, or other liquid, flows through a narrow tube, the velocity of flow will depend on the nature of the liquid and the force which produces the flow. All parts of the liquid, however, do not move through the tube with the same velocity. A very thin layer in contact with the wall of the tube remains stationary, and over a short distance the successive very thin layers of liquid move over one another with gradually increasing velocity, until a maximum is reached. There is thus a shearing or a movement of the different layers past one another in the direction of flow; and this displacement of the different layers relatively to one another is opposed by the *internal friction* or *viscosity* of the liquid.

When the tube is a narrow one, therefore, one can regard the column of liquid as made up of a number of concentric tubes sliding past one another like the draw-tubes of a telescope; and the resistance to the flow is due to the friction between the different layers of liquid. In the case of a fairly mobile liquid the zone of variable rate of flow in the neighbourhood of the walls of the tube is a narrow one and outside this zone, in the middle of the tube, the liquid streams, without friction, at a constant rate.

When the liquid is moving through a narrow tube there will be a constant difference in velocity between the different tubes of which we have regarded the cylinder of liquid made up, and it has been found that the force per area which is necessary to maintain this condition is proportional to the difference of velocity v of two adjacent tubes (or their relative velocity of displacement) and inversely proportional to their distance x apart, *i.e.,* where 77 is a constant known as the *coefficient of viscosity.* When the velocity of displacement of two layers is numerically equal to the distance between the layers *(v~ x),* the force in dynes per unit area becomes equal to the coefficient of viscosity. This gives the definition of the latter quantity. The reciprocal of the viscosity, or 1/77, is called the *fluidity.* The unit of viscosity in C.G.S. units is known as a *poise* (Poiseuille).

For the flow of a homogeneous liquid through a capillary tube the following expression (Poiseuille's formula) has been deduced: where p is the driving force, r is the radius of the tube, t is the time required for the volume V of liquid to flow through the tube of length *I.* This formula holds strictly, however, only when the velocity of the liquid on leaving the tube is zero; and when this is not the case, a correction for the kinetic energy of the liquid must be introduced.

The apparatus now generally employed for the determination of the viscosity of liquids is the Ostwald modification of Poiseuille's apparatus.

SURFACE TENSION

When a liquid is withdrawn from the action of gravity, as, for example, when it is suspended in a medium of the same density, it assumes the form of a sphere; and whenever a free surface is formed the liquid tends to form the smallest possible surface. This behaviour is due to the fact that an attractive force exists between the molecules of a liquid whereby an " internal pressure " is produced.

A molecule in the body of a liquid is equally attracted in all directions, but a molecule in the surface layer is subject to an unbalanced force acting inwards at right angles to the surface. The surface of a liquid, therefore, may be regarded as being the seat of a special force, the action of which can be visualised by imagining the surface layer of liquid to act as an elastic skin which tends to contract.

This force, acting perpendicularly to a section of the surface 1 cm. long, is called the *surface tension,* and is generally expressed in dynes per centimetre. Usually one thinks of the surface tension as the force existing at the interface between a liquid and its vapour or between a liquid and air, but a surface tension also exists at the boundary surface between two immiscible liquids, between a liquid and a solid, and, indeed, whenever a free surface is formed. The value of the surface tension will depend on the nature of the interface.

When a liquid wets a solid the surface tension is able to do work, as when water rises in a glass tube or in a strip of blotting paper.

If, however, the liquid does not wet the solid, the surface tension will resist the doing of work —or a certain amount of work must be done in order to break through the surface layer. Thus, when a glass tube is placed in mercury, a certain amount of work must be done in order to depress the tube through the surface of the mercury, and the level of the mercury inside the tube is lower than the level outside.

The existence of a surface tension and the fact that the surface layer of a liquid acts as if it were an elastic skin, afford an explanation of a number of familiar phenomena. If water or a water paint is brushed over a piece of greasy paper, it does not spread evenly as a film, but forms into drops owing to the action of surface tension, just as mercury forms drops or globules on a surface which it does not wet. If, however, the surface tension of the water is reduced, that is. if the imaginary surface skin is weakened, then the water will spread out as a film over the paper. The lowering of the surface tension of the water can be effected by the addition of various substances, *e.g.,* sodium soaps. Further, it is owing to the existence of a special surface layer that it is possible to float a steel needle or a boat of wire gauze on the surface of water, or to carry water in a vessel of wire gauze. If the needle or wire gauze be coated with a film of grease or of paraffin, so that it is not wetted by the water, the needle or gauze will merely form a depression, or series of depressions, on the surface of the water, as if resting on a thin elastic film.As was pointed out, if a liquid does not wet a solid, a certain amount of work has to be done in order to break through the surface layer of the liquid.

PROPERTIES OF CRYSTALLINE SOLIDS

If one were asked for a definition of the solid state, one would probably quickly answer that a solid is a rigid body which does not flow and which possesses a definite shape and volume. Such a definition, however, is soon found to be unsatisfactory, for the property of flowing is shown by many bodies, *e.g.*, sealing wax, which one would certainly class as solids. Moreover, even although, generally speaking, a solid differs from a liquid in possessing the property of *elasticity, i.e.*, of recovering from a deformation when the deforming force is removed, this property also is found to be too indefinite to separate sharply solids from liquids.

The classification, indeed, of bodies into solids and liquids is itself not a satisfactory one, for *amorphous solids,* on being heated, gradually lose their rigidity and elasticity and pass *continuously,* without any sudden change of properties, into liquids. Amorphous solids, therefore, cannot be sharply differentiated from liquids.

Between *crystalline* solids and liquids, however, a sharper distinction can be drawn. In the case of crystals, the structural units (atoms, molecules, or ions) exist in definite and orderly array *in a space lattice*; and the orderliness of internal arrangement makes itself manifest in an external orderliness of geometrical forms which are bounded by faces inclined to one another at definite angles. In the case of liquids, however, even when the atomic or molecular units are not distributed entirely at random, there is a breaking down of the space lattice and a loosening of the intermolecular forces which maintain the rigidity of a crystal. There is, moreover, no continuous change from the crystalline to the liquid state, for a crystalline solid has a sharply defined *melting-point* at which it undergoes an abrupt change into a liquid.

In an amorphous solid the physical properties (*e.g.*, thermal expansion, elasticity, solubility) have the same values in all directions. Amorphous solids are *isotropic.* In crystals, however, except in the case of those belonging to the cubic system, the physical properties have different values according to the direction. Such crystals are said to be *anisotropic.* If a sphere of glass, for example, is heated, expansion takes place equally in all directions and the spherical form persists; but if a sphere of, say, rock crystal, is heated, expansion is the same in all directions and an ellipsoid is obtained.

If the polished surface of a section through a crystalline substance is treated for a short time with a solvent, a pattern may be produced owing to the unequal rate of solution in different directions. The etch figures, as they are called, are all alike on similar faces but unlike on dissimilar faces. They conform to the symmetry of the class to which the crystal belongs, and by examining them by means of a microscope the crystal form of the substance may be determined. The production of such etch figures is of great importance in the investigation of metals and their alloys.

FORMATION OF CRYSTALS

As the temperature of a liquid is lowered, the forces of attraction between the molecules gain more and more the upper hand over the disrupting and separating action of molecular motion; and under favourable conditions molecule joins itself to molecule, not vaguely and uncertainly in loose attachment but in a firm linking together at definite points under the action of van der Waals forces. Molecular motion still exists but is restricted to a vibration or oscillation about certain fixed points. In this way, out of the general welter, small groups of molecules are formed in definite array. Around each such invisible group of molecules other molecules marshal themselves, and the structure grows into the visible crystal.

Crystallisation also may take place from solution. In an ordinary solution the molecules of the solute are kept apart by the molecules of the solvent; but as the solution is concentrated by evaporation, or as the temperature of a concentrated solution falls, the molecules of the solute become more crowded, or their kinetic energy diminishes to such an extent that the molecules are able to join up into crystal units, to which more and more molecules attach themselves in orderly array and so build up the visible crystal.

The structural units of crystals may not only be *molecules,* bound together by van der Waals' forces, but also *atoms (e.g.,* diamond), linked by covalent bonds, or *ions (e.g.,* sodium chloride), held together by electrostatic forces. In the case of metals, the structural units are positive ions with free electrons. The number of free electrons in unit volume varies with the temperature. For the most part, metals are close-packed assemblages of cubic, body-centred cubic, or hexagonal unit cells, and the bonds between the atoms are spatially undirected. Deformation by gliding can thus readily take place giving rise to the properties of malleability and ductility.

The production of a perfectly formed crystal, however, requires the exercise of great care and the existence of favourable conditions. Thus the growth of the crystal must be slow in order that the molecules may have time to arrange themselves properly; and the crystal must also be allowed to grow freely in all directions. When these conditions do not exist the crystal becomes distorted in various ways, and great divergence from the form of the perfect crystal may be observed. The external " habit " of the crystal may vary greatly.

Thus when a thin layer of liquid crystallises on a flat plate, growth can take place practically in only two dimensions, and flat fan-like or fern-like crystals are formed. This is observed, for example, when moisture freezes on the window-pane. Such distorted crystals are called *dendrites,* or dendritic crystals, from their resemblance to tree growths. The beautiful leaf-like or lace-like crystals seen in snow-flakes have been formed owing to crystalline growth having been more rapid in certain directions than in others. It may, however, be noted that although the " habit " of a crystal may vary, the crystallographic system to which the crystal belongs remains the same.

Presence of impurities also may produce a variation in the habit of a crystal. Thus sodium chloride forms " hopper crystals " when allowed to crystallise from a solution con taining a small amount of impurity (alum). This behaviour finds an important practical application in industry for the production of needle-shaped crystals of ammonium sulphate which do not cake together.

Not only may molecules link themselves together into a crystalline structure but they may sometimes link themselves into two or more different kinds of structure. This is the phenomenon of *polymorphism.* Thus sulphur forms not only the well-known rhombic crystals but also needle-shaped monoclinic crystals; and mercuric iodide crystallises in two different forms, one of which is scarlet and the other yellow in colour.

MELTING-POINT AND FREEZING-POINT

When a crystalline solid is heated it passes at a certain definite temperature into a liquid; it melts or fuses. The temperature of the mixture of solid and liquid, moreover, remains constant, on further addition of heat, until all the solid has melted, owing to the fact that heat becomes latent when a crystalline solid melts. This is known as the *latent heat of fusion. At the melting-point the solid and liquid are in equilibrium.*

Since a crystalline solid melts at a definite temperature, it might also be expected that when a liquid is cooled down to the melting-point of the solid the liquid would begin to crystallise. In most cases, however, this does not happen. One finds, on the contrary, that it is frequently, and indeed generally, possible to cool a liquid considerably below the temperature at which the solid melts without any crystallisation taking place. The liquid is then said to be *supercooled.* In order that crystallisation shall take place a "centre of growth " or a crystalline " nucleus " must first be formed round which the molecules may arrange themselves.

In most cases, if the cooling is carried out sufficiently slowly, nuclei or stable crystal units are formed spontaneously by the accidental meeting, it may be, of the molecules in a suitable way; and vigourous stirring assists these favourable encounters. The number of nuclei formed increases with the degree of supercooling, passes through a maximum, and then decreases. In some cases, however, spontaneous formation of nuclei may not take place, and it may be necessary to introduce a small crystal from the outside in order to start the process of crystallisation.

For this purpose the tiniest particle (say, one-millionth of a gram) is sufficient. It would appear, therefore, that the *freezing-point* of a liquid, unlike the melting-point of a crystalline solid, is rather indefinite. If, however, one defines the freezing-point as the *temperature at which the liquid is in equilibrium with the crystalline solid,* the temperature becomes quite definite and identical with the melting-point.

It should be noted that crystallisation in a supercooled liquid or in a supersaturated solution cannot be initiated by any chance solid but only by a crystal of the fused or dissolved substance or by a crystal of an isomorphous substance. Thus a crystal of sodium acetate will not initiate crystallisation in fused sodium thiosulphate, but if a crystal of chrome alum is introduced into a supersaturated solution of potash alum the latter will crystallise out on the former, since the two substances are isomorphous.

Chapter 15

Industrial Chemistry

Industrial chemistry deals with the preparation of products from raw materials, through the agency of chemical change But there is an occasional exception to this definition, for a few industries, depending on strictly mechanical changes, are classed among the chemical industries Since a sharp line cannot be drawn between chemical and mechanical technology, a study of the former necessarily involves some consideration of the mechanical appliances and apparatus, by means of which the chemical reactions are carried out.

The products of chemical industry are exceedingly numerous and varied in character, but comparatively few come into the hands of the mass of the people for direct consumption Many of them are used only in making other substances, for it is often the case that the finished product, by-product, or waste from one industry becomes the raw material for another, and it rarely happens that one manufacturer, starting with the raw materials found in nature, produces from them articles for popular use. Thus the chemical industries become a network of interlacing processes, and in considering one it is often difficult to separate it from others which have a more or less direct bearing upon it. Furthermore, as competition has become very close in many lines, the use which may be made of by-products and waste is so important, that processes are often carried out with the view of obtaining larger yields or better quality of the by-products, which may have become a source of considerable profit In a few instances, it might be said that what were originally the by-products are now the chief products and main support of these particular industries. This is especially true in the case of the Leblanc Soda Industry, which would long since have been abandoned were it not for its production of hydrochlonc acid The utilization of waste materials furnishes an almost inexhaustible subject for investigation by the industrial chemist.

The manipulations of most frequent occurrence in the various processes are here defined and explained for the sake of brevity in the text.

LIXIVIATION

Lixiviation is the process of separating soluble from insoluble substances by dissolving the former in water or some other solvent. The mixture of substances is put into a suitable vessel, the solvent poured over it, and the whole allowed to stand until a strong solution is obtained, which is then drawn off from the residue. This process is repeated as often as necessary, until the desired amount of soluble matter has been removed Sometimes the mixture is put into baskets, or on gratings, which are suspended in tanks of water The solution being denser than the solvent sinks to the bottom as it forms, and water comparatively free from dissolved material is thus constantly brought into contact with the substance to be lixiviated The insoluble substance remains on the grating or in the baskets When desired, the soluble material may be recovered from the solution by evaporation or precipitation Extraction is the term usually employed when some solvent other than water is used in lixiviating. Thus we speak of extraction by steam, alcohol, carbon disulphide, etc.

LEVIGATION

Levigation is the process of grinding an insoluble substance to a fine powder, while wet. The material is introduced into the mill together with water, in which the powdered substance remains suspended, and flows from the mill as a turbid liquid or thin paste, according to the amount of water employed. There is no loss of material as dust, nor injury or annoyance to the workmen Further, any soluble impurities in the substance are dissolved* and the product thereby purified. The greatest advantage of this process is the facility it affords for the subsequent separation of the product into various grades of fineness, because of the slower subsidence of the finer particles from suspension The turbid liquid flows into the first of a series of tanks, and is allowed to stand for a certain time The coarsest and heaviest particles quickly subside, leaving the finer material suspended in the water, which is drawn from above the sediment into the next tank The liquid is passed from tank to tank, remaining in each longer than it remained in the preceding, since the finer and lighter the particles, the more time is necessary for their deposition In some cases a dozen or more tanks may be used, and the process then becomes exceedingly slow, as very fine slimes or muds may require several weeks for the final settling But as a rule, from three to five days is sufficient.

The term " levigation " is now often applied to mere sedimentation, a substance being simply stirred up in water, without pre\ lous wet-grinding, in order to separate the finer from the coarser particles, as above.

EVAPORATION

Evaporation, in a technical sense, denotes the conversion of a liquid into a vapour for the purpose of separating it from another liquid of higher boiling point, or from a solid which is dissolved in it In the great majority of cases, the liquid evaporated is water If the liquid evaporated is to be recovered, the vapors are condensed, and the process then becomes one of Distillation.

There are four general methods of evaporation:

1. Spontaneous evaporation in the open air
2. Evaporation by application of heat directly from a fire to the vessel containing the liquid

3. Evaporation by indirect application of heat from the fire, as by means of steam, with or without pressure
4. Evaporation under reduced pressure

The first method, by spontaneous evaporation in the open air, is comparatively slow, and requires exposure of very large surfaces of liquid The time necessary depends upon the temperature and humidity of the air, and the completeness with which the vapors are removed from the surface of the liquid, hot, dry weather, especially if a brisk wind is blowing, evaporates water quite rapidly. This process is only used for the manufacture of salt from sea water, or from natural brines. In certain warm countries considerable quantities of salt are thus prepared, and in this country some is made from a brine found near Syracuse, N Y. Sometimes weak brines are allowed to trickle in fine streams over tall piles or "ricks " of brushwood in the open air. The liquid being so exposed in thin layers, to the air and wind, is concentrated to such a degree that it will pay to complete the evaporation by artificial heat.

The second method, by direct application of heat from a fire, is very largely used in the arts. This may be done in two general ways.

a. The flame?, or hot gases from the fire, are generally allowed to play directly on the bottom of the vessel containing the liquid; or they may pass through flues or pipes, set into the vessel, so that the liquid surrounds them on all sides. Such pans are often several yards in length, and may contain one large flue, or several small ones, according to the work desired; but this form of apparatus is expensive to build, and difficult to keep in repair.
b. The flames and hot gases may be conducted over the surface of the liquid to be evaporated. This mode is only used for coarse and common products, or in the concentration or recovery of waste materials But it has the advantage that the bottom of the pan is less liable to be injured by the crusting of a precipitate upon it Another point often in favour of surface

heating is that the liquid is evaporated in a reducing atmosphere But as flue dust and ashes are liable to fall into the pans, the product is usually impure Large shallow pans are used, which are generally arched over with brick, in order that the heat may be better utilized, through radiation from the brick walls There are various ways of setting the pans for this process. A modification of this method is the use of a long cylinder, set at a slight incline, and revolving about its longitudinal axis. The lower end is open for the entrance of the flames and gases from the grate, which pass through the cylinder, on their way to the chimney. The hot gases are often passed through the flues of a boiler, to utilize the waste heat. The solution to be evaporated is fed into the cylinder at the upper end in a small stream, and comes in direct contact with the flame. The water is evaporated, and the solid matter is delivered into the pit or wagon at the lower end of the furnace, in a dry and calcined state. Such furnaces are frequently used for evaporating waste liquors to recover the salts which they contain, and for the treatment of sewage and other liquid refuse.

The third method of evaporation, by the use of steam heat, is very often employed where there is danger of injury to the product by overheating

a. Jacketed pans or kettles may be used. These are simply double-walled vessels, the steam being admitted between the walls

b. The steam may be allowed to circulate through coils of pipe, placed inside the vessel, which is sometimes made of wood. The temperature of the liquid depends on the steam pressure, very often exhaust steam is employed

The fourth method, evaporation in vacuo, is merely a modification of either the second or third method, but is considered separately for convenience. The boiling point of a liquid may be very materially lowered by reducing the pressure within the vessel. Hence, solutions containing substances which would be injured by the heat necessary to boil them under the atmospheric pressure, or liquids boiling at very high temperatures, are evaporated in vacuum pans.

The different forms of apparatus used for vacuum evaporation vary much in their details, but all depend on the principle of reduced pressure.The essential parts of the plant are the vacuum pan or still, the pump for exhausting the air and steam from the pan and sending them to the condenser, and the heating apparatus. The vacuum pan is usually a globular copper or iron vessel, provided with a manhole, a pressure gauge, and a discharging valve. Very often a piece of heavy plate glass is set in the side to afford a view of the interior during evaporation. On the top of the pan is a dome or short tower, from which a pipe leads to a receptacle, called the catch-all," that retains any liquid which may escape from the pan. A small pipe returns this liquid to the pan, and a larger one connects the " catch-all " with the vacuum pump, which is an ordinary double-cylinder air pump of large size, driven by an engine An injector pump, which condenses the steam directly, may be used. The pan is generally heated by steam coils within it, or by a steam jacket, or by both.

A very efficient method of vacuum evaporation is that obtained by the use of Multiple Effect Systems In these greater economy of fuel for heating is secured The apparatus consists usually of two or more simple vacuum pans, so joined together that the steam from the boiling liquid in the first pan is made to pass through the coils and jacket of the second pan, and the steam generated in the second pan goes through the coils and jacket of the third, and so on through the system. The vacuum maintained in each pan of the series is greater than in the one preceding Hence, notwithstanding its increased concentration, the boiling point of the liquid in the second pan is so low, that the steam from the first pan is sufficiently hot to boil it Similarly the steam from the second pan is made to boil the liquid in the third, in which there is still less pressure, and so on to the last pan, in which the highest vacuum is maintained

As a rule only four pans are used, for it is very difficult to sustain the vacuum sufficiently to work another pan in the series In many plants only three pans (triple effects) are used

An effective modification of this method is the apparatus known as the Yaryan evaporator It is made in triple and quadruple effects, and each pan is exactly like its neighbors It consists of an outside shell of iron, within which is a system of small tubes, joined together in groups of five or six, each group constituting a section or unit The tubes in each unit are so connected at the ends as to form one continuous coil The liquor to be evaporated is run through the several coils thus constructed in each pan. The tubes in the first pan are heated by steam, introduced into the shell directly from a boiler As the liquid flows through the tubes, it is brought to boiling, and the steam generated mingles with it, converting the whole mass into foam, which runs through the coil and spurts against a baffle plate in the " separator", which is an enlarged chamber at the end of the shell. The steam and liquid are separated, the liquid falling to the bottom and running off into the receiver, to be passed through the tubes of the next pan The steam rises, passing through the steam dome and "catch-all", and into the shell of the next " effect," through the coils of which the liquid is passing under still greater vacuum, and so on through the system. The apparatus is very economical in its use of fuel, and as the liquid is exposed in thin films to the heat, the evaporation is rapid, hence the liquid is subjected to a high temperature for only a short time.

The apparatus is nearly automatic in its action, and needs little attention It can be stopped and started quickly, since it contains only a small quantity of liquid at one time, and it occupies but little floor space when the several " effects " are placed one over the other

The ordinary form of vacuum pan evaporates about 8 lbs of water per pound of coal, but it is said that the best forms of Yaryan apparatus evaporate from 23J to 25 lbs of water per pound of coal in a triple effect, and 30 lbs in a quadruple effect.

The Kestner evaporator utilizes the " climbing film " principle, by which the liquid to be evaporated is automatically distributed over the heating surface, without the use of pumps The apparatus is built in multiple effect and each pan consists of a narrow vertical shell containing a number of small tubes through which the liquor passes upward from the bottom The tubes are tightly fixed into plates at top and bottom and are entirely surrounded and heated by steam in the shell. The tubes are about twenty-three feet long and are open at top and bottom Liquor is fed in through the valve and the supply so limited that a relatively small quantity enters each tube, where it at once begins to boil:

the vapour evolved rușhes up the tube, carrying some of the liquor along and distributing it in a thin film on the hot tube wall. Emerging from the top of the tubes the foaming mixture of vapour and liquor is discharged against the vanes of a centrifugal separator by which the concentrated liquor is whirled against the walls of the enlarged vapour space The liquid flows down the walls and passes out at, while the vapour rises and passes to the shell of the next effect, or to the condenser The separation of vapour and liquor is claimed to be so complete, that only a small quantity of liquid is in the apparatus and the time of heating is short, only about two minutes being required for the passage through the tubes The drop in temperature between any two pans of the series is small, ranging from 8° to 12° C Thus four and five effects are often used in series The apparatus occupies little floor space, gives little trouble from scaling of the tubes, and is easily washed out.

The Lillie evaporator is a very efficient type of multiple effects. Slightly inclined straight tubes tightly fastened at one end in the thick plate open into the steam space. The other ends of the tubes are closed, except for a small air vent, and are unsupported Thus they expand and contract freely, preventing strains and resulting leaks. In the upper part of the effect is a row of distributing pipes, each having a longitudinal slot on its upper side.

These pipes are closed at one end; the other opens into a distributing box The liquor to be evaporated enters through, passes into, and flowing from the slots in thin films, is showered uniformly over the hot tubes, from whose outer surface the evaporation takes place The liquor drips from tube to tube, collecting in the float box, from which the suction pipe of the centrifugal pump draws it, to again pass over the tubes The float in the box operates a valve which allows fresh liquor to enter the effect just fast enough to replace that vaporized and what passes from the discharge pipe as concentrated liquor On is a regulating valve governing the level of the liquor in and thus controlling the rate of feed, the slower the discharge, the greater the concentration The float completely closes the feed valve when the liquor rises to a definite height in, the discharge valve in the last effect thus automatically controls the flow of the liquor from effect to effect, by influencing the action of the feed valves The tubes are heated by live or exhaust steam, or by vapour from the preceding effect, which enters the steam chamber, the hot water condensed in collects in the bottom of, and passing the steam trap goes to the steam space of the next effect, thus being under great vacuum, it gives up part of its heat as steam, which assists in the heating of this effect.

DISTILLATION

Distillation is the process of vaporizing a liquid and recovering it by condensing the vapors.The liquid formed by this condensation is called the distillate. Distillation is chiefly employed to separate a liquid from non-volatile matter dissolved or suspended in it, or to separate one liquid from a mixture of liquids of different boiling points, that one having the lowest boiling point being the first to begin to pass off as vapour

The separation of two miscible liquids by distillation depends on the difference between the composition of the vapour and of the boiling liquid from which it comes, and while never perfect is more complete the greater the difference in composition Most liquid mixtures evolve a vapour containing more of the low-boiling constituent than does the liquid itself, but in some cases the reverse is true between certain limits of composition, and such liquids always give mixtures of maximum or minimum boiling point, which cannot be separated by distillation as the vapour and liquid compositions are in these cases the same.

During a distillation the boiling point gradually rises, and at the end there remains in the still a relatively small amount of the high-boiling liquid very free from the other component, or else a mixture of maximum boiling point While the distillate has been enriched in the low-boiling constituent, it is far from pure, but repetition of the distillation improves the separation In general it is easier to secure the high-boiling liquid free from the other than the reverse.

If a mixed vapour be slowly cooled, the liquid to condense is that in equilibrium with the vapour, thus it is largely the less volatile component which separates first By abstracting only enough heat to condense a part of the vapour, the remainder is greatly enriched in the volatile constituent, this fractional condensation is attained by using a condenser with relatively hot cooling medium, the uncondensed vapors passing to a cold condenser for complete condensation. From the fractional or partial condenser, the condensate returns to the still for reboiling, to remove the remainder of the volatile component Fractional condensation is equivalent to a redistillation, without the consumption of additional heat.

The chief parts of every distilling apparatus are the boiler or still and the condenser The still is usually iron or copper, and may be heated directly by a furnace, or by a steam-jacket, or a coil The condenser is a coil of pipe, or a system of tubes, or a double-walled chamber, submerged in a tank of cold water Condensers are usually made of iron or copper, but lead, silver, earthenware, or glass tubes are sometimes used The fractional condensation apparatus is placed between the still and the final condenser, it may consist of a series of chambers, or of pipes or U-tubes, surrounded by a water-bath or other liquid at a temperature between the boiling points of the liquids to be distilled.

The condensate from a fractional condenser is richer in the volatile component than is the liquid in the still, it is, therefore, not in equilibrium with the vapour from the still, and if brought into contact with that vapour, an interchange of heat results, accompanied by a volatilization of the low-boiling constituent and a condensation of the high-boiling one, which further enhances the vapour and improves the separation, without any additional expenditure of heat This process is called dephlegmation, and to secure the maximum efficiency, must be repeated a number of times The mixed vapors from the still are bubbled through a series of shallow layers of liquid, through which passes in counter-current flow, the partial condensate from the fractional condenser, this condensate, with its low-boiling constituent removed as completely as possible, finally flows back into the still.

The apparatus for this, called the dephlegmator, or " column usually consists of a tall tower set above the still, and divided by a number of perforated horizontal plates into many shallow chambers, through each of which the returning liquid must flow on its way to the still The vapors from the still pass up through the openings in the plates, bubbling through the shallow layers of the liquid and give up heat to it, causing vaporization of its volatile constituents The boiling of the liquid in all of the chambers is done by the heat in the vapors entering the lowest chamber any condensation in the tower results in greater coal consumption to restore the heat thus lost The chief object of the dephlegmator is to boil the liquid in the several chambers, and deliver the enriched mixed vapors to the fractional condenser As radiation from the walls of the column causes condensation and loss of heat, the column should be covered or lagged.

In an ordinary distillation, even with use of fractional condenser and dephlegmating column, it is only at the end of the operation the there remains in the still the high-boiling component in a practically pure state, at any earlier stage in the process, however, it is possible to dephlegmate the low-boiling component from the liquid in the still, by using a dephlegmating column below the still This is done in practice by admitting the liquid mixture to be distilled into the middle of a long column, the overflow from which passes down to the still in which there is only pure high-boiling liquid, the rising vapors carry the low-boiling constituent The original mixture can now be admitted and the distillate and high-boiling residue withdrawn continuously from the apparatus Also, the column, below the point of admission of the liquid to be distilled, accomplishes the redistillation of the overflow from the apparatus, without further consumption of heat, in addition to the great advantage of continuity of operation, this type of column gives the best heat economy of any method of distillation, and is generally employed except where special conditions render it inadvisable. The heat economy can be still further improved if the feed liquor be heated by using it as cooling medium in the fractional or final condensers.

ORGANIC INDUSTRIES

DESTRUCTIVE DISTILLATION OF WOOD

Wood consists mainly of cellulose with its incrusting layer, and of sap, containing water, resins, tannins, colouring matter, and mineral salts wood contains 15 to 20 per cent of moisture When heated in closed retorts, away from the air, the cellulose and ligneous matter decompose, after the moisture is expelled, and a complex reactions occurs, by which a great number of substances are formed The crude products are gases, thin liquids, viscous liquids or tar, and charcoal. When wood is carbonized in pits, the volatile products go to waste, by the use of retorts, the valuable liquid distillates and tar are saved, the gases evolved are mainly hydrogen, methane, ethane, ethylene, carbon monoxide, and carbon dioxide, they have no value for illuminating, and are burned under the retorts, thus economizing fuel.

When wood is heated in retorts, the moisture is driven out, but no decomposition occurs until the temperature approaches 160° C., between 160° and 175°C,a thin, watery distillate, called " pyroligneous acid," is formed, above 275° C, the yield of gaseous products becomes marked, and between 350° and 450° C, liquid and solid hydrocarbons are principally formed Above this last temperature, little change occurs, and charcoal, containing the mineral ash, remains in the retort.

The pyroligneous acid contains the important distillates, methyl alcohol and acetic acid, together with acetone, methyl acetate, phenols, ketones, and other substances The tar contains aromatic hydrocarbons and paraffines Its most valuable constituent is the creosote oil, containing guaiacol, creosol, and other phenols of high molecular weight. A comparatively small amount of phenol or carbolic acid is present, however.

The proportion of gaseous products to liquid distillate and charcoal is affected by the method of heating; rapid heating to a high temperature increases the quantity of gas, by distillation at a low temperature, the yield of pyroligneous acid, tar, and charcoal is larger. The wood used affects the amount of acid and tar, deciduous trees, especially birch, oak, and beech, are preferred, coniferous woods yield less acid, but afford a tar containing much resin and turpentine The yield of acid and tar is increased by the rapid removal of the vapors from the retort.

Wood is distilled in various kinds of kilns or retorts. If charcoal is the only product in view, the carbonization may be done in " pits " or kilns, and the volatile products go to waste Masonry kilns of large capacity (15 to 90 cords at a charge) are often used, the necessary heat may be derived from combustion of part of the charge itself, or from an external fire whose combustion gases pass into the kiln, or through flues in its walls.

When the volatile products are to be saved boiler-plate iron retorts, externally heated, are employed, these may be either stationary or movable, and of large or small capacity Small horizontal cylinders, holding about 1 or 1J cords at a charge, are usually set two in a bench, over a common furnace Each retort connects with a separate condenser by the copper pipe, a pipe carries the uncondensed gases to the grate where they are burned under the retorts The wood, cut to proper length, is rapidly filled into the retort, which is still hot from the previous charge, the door is closed and distillation begins at once After some 12 hours, the hot charcoal is rapidly drawn into an iron box to cool out of contact with the air, and the retort at once recharged.

For larger output, horizontal oven-retorts are used. These are rectangular iron boxes, into which several steel cars, each loaded with 2 to 4 cords of wood, are run at one time, and the carbonization carried on for 24 hours Thus charges of 10 to 20 cords of wood at a time are expeditiously handled When carbonization is finished, the doors at each end of the retort are opened, and a string of newly loaded cars pushed in, which also shoves out the cars carrying the hot charcoal from the previous charge, into a large iron box or cooler, placed directly opposite, where they cool out of contact with the air This transfer and introduction of a new charge requires only a few minutes, so these retorts are practically continuous in action and the loss of charcoal by combustion is small This system consumes less fuel, and has lower labour and repair costs for a given output than the small retorts.

Movable retorts are vertical boiler-plate cylinders so arranged that the retort, filled with wood and with cover on, is lowered by a crane into the furnace, when carbonization is finished, the retort containing the charcoal is lifted out to cool unopened, while another charged with wood is put into its place Connection with the condenser is made by a copper swing-pipe, clamped to the vent on the top of the retort Each cylinder holds one cord of wood, and as it is packed cold, a complete filling of the space is possible The labour cost for this type is less than for small horizontal retorts, but the wear and tear on the furnace and retorts from the frequent moving and cooling is great; the original cost of plant is also higher.

Coniferous woods are often distilled in retorts into which superheated or free steam is introduced, preliminary to the distillation proper. Thus the turpentine and much rosin are driven out at temperatures below that at which the cellulose is decomposed. Then by raising the heat of the retort, true destructive distillation follows, yielding wood vinegar, tar, and charcoal Many plants for distilling fat pine, " light wood," have been erected in this country, but not always with satisfactory results. Considerable wood turpentine is produced, the yield being 10 to 20 gallons of crude turpentine per cord of " light wood but the odor and colour are often not satisfactory.

Extraction processes, depending on the use of solvents for both turpentine and rosin, such as carbon disulphide or carbon tetrachlonde or turpentine itself, have been patented The extracted chips are steamed to recover the solvent, but the losses are so large that the future of these methods is uncertain The extracted chips are suitable for wood pulp, or they may be distilled for pyroligneous acid and tar. By treating the chipped wood with caustic soda liquor and steaming, the turpentine can be distilled off, while the rosin dissolves in the alkaline liquor. When drawn off and acidified, the rosin is precipitated.

Pyroligneous acid or crude " wood vinegar " is a reddish brown liquid with strong acid reaction and empyreumatic odor due to fur-furol in part. It averages 5 to 10 per cent acetic acid, 1 5 to 3 per cent methyl alcohol, and 0 1 to 0 2 per cent acetone, its specific gravity varies from 1 020 to 1.050 A small amount is used directly for making an impure iron acetate, sold as " pyrolignite of iron ", but it is usually worked for methyl alcohol and acetic acid.

By neutralizing pyroligneous acid directly with milk of lime and distilling, a raw wood spirit is collected as distillate, while calcium acetate solution remains in the still, which on evaporation to dryness yields so-called " brown acetate of lime," averaging about 67 per cent of calcium acetate Dunng the evaporation, tar separates as a scum, which is skimmed off The method is not much used at present.

By distilling pyroligneous acid in a copper still, the tar is left and a purified " wood vinegar " obtained, containing acetic acid, methyl alcohol (wood spirit), methyl acetate, acetone, acetaldehyde, etc, and traces of tar, empyreumatic matter, etc. This is neutralized with lime, precipitating many of the impurities; the clarified solution is then rectified in a column still, yielding wood alcohol of about 82 per cent. The solution in the still yields on evaporation to dryness "gray acetate of lime," averaging 80 per cent calcium acetate The tarry matter decomposes during the drying.

The raw wood alcohol is further purified by diluting with water until the oily matters (ketones, aldehydes, etc.) precipitate, as shown by the milky appearance. On standing several days, the oils rise to the top and are skimmed off, then the alcoholic solution is again distilled in a fractionating still, until the concentration is about 95 per cent alcohol, the product is known as wood spirit or methyl alcohol By filtering through charcoal, the colour and unpleasant odor can be largely removed. Treatment with caustic lime and redistillation yields alcohol of 99 per cent, or higher, concentration, but this does not remove the acetone.

The wood spint may be purified from acetone by treatment with caustic soda and iodine, producing a precipitate of lodoform with the acetone Or calcium chloride is added to combine with the methyl alcohol and form a crystallized solid, stable at 100° C, from which the acetone is distilled off, then by adding hot water and heating to 100° C, or over, the alcohol is distilled off and rectified.

Commercial methyl alcohol is often slightly yellowish in colour and frequently has a disagreeable odor. It is much used as a solvent in varnish making, for which purpose the presence of acetone is desirable, for making formaldehyde, for mixing with ethyl alcohol to prepare " denatured ethyl alcohol " or " methylated spirit ". Acetone, when recovered from wood spint, is generally distilled from the calcium chlonde compound with the methyl alcohol It is, however, more commonly made by the dry distillation at 290° to 300° C., of calcium acetate:

$$Ca(C_2H_3O_2)2 = CaCO_3 + CH_3 + CO + CH_3$$

The product obtained by either of the above methods is crude, sodium bisulphite is added, and forms a double salt with the acetone, which is purified by recrystallization from aqueous solution.

This salt is decomposed by heating with sodium carbonate solution, liberating the acetone, which distils off pure Or the crude acetone is neutralized with lime, settled, and the supernatant liquid diluted with water and rectified in a column still, yielding a pure acetone distillate and an oily residuum called acetone oil which finds some use for solvent and denaturizing purposes.

A method for making acetone devised by Dr. E. R. Squibb consists in passing acetic acid vapour through a rotating iron cylinder, heated to about 500°-600° C, and containing pumice stone with precipitated barium carbonate. The barium carbonate acts as a contact body, since the temperature is always above that at which barium acetate decomposes The vapors from the still pass to a fractional condenser to remove water and acetic acid, the acetone condenses in a second condenser.

Acetone is a colourless mobile liquid, having a peculiar odor and unpleasant taste, its specific gravity is 0 797 at 15° C and should not exceed 0 802 in the commercial product It boils at 56 3° C, and mixes with water in all proportions, is an excellent solvent for many resins, gums, fats, nitrated cellulose, and other substances It is used in the manufacture of celluloid, smokeless powders, chloroform, iodo-form, sulphonal, for extracting resin from crude rubber, and for denaturizing ethyl alcohol.

Commercial acetic acid is prepared from gray or brown acetate of lime by distilling with strong hydrochloric or sulphuric acid. In the hydrochloric acid process, copper stills heated by steam coils are used, free steam can also be blown into the charge The acetic acid distils over, leaving calcium chloride in the still. The acid is a slightly coloured liquid containing from 30 to 50 per cent of anhydrous acid, according to the strength of the hydrochloric acid used It may be further purified by distilling again over a little potassium permanganate, and filtering through fresh charcoal.

The sulphuric acid process is more commonly used at present, the still is made of cast-iron with a scraping device inside. The distillate contains about 75 per cent anhydrous acetic acid and usually a little sulphur dioxide. This acid is then rectified in a large copper column still, heated by a steam coil. The plates in the column are often porcelain or earthenware, the condensers are usually copper worms.

If air is excluded from the apparatus, there is little attack on the copper by the acid, but this necessitates the immediate refilling of the still after each charge has been worked off. If the operation is to be discontinued, the still and condensers should be thoroughly washed out with water When starting each distillation, the heating should be slow, to allow the sulphur dioxide to pass off before the acetic acid begins to distil.

According to the amount of cooling water admitted to the fractional condenser, a clear colourless liquid containing from 80 to 99 per cent of anhydrous acid can be obtained, it contains traces of empyreumatic matter, which can be removed by gentle heating with potassium permanganate in an earthenware vessel, and redistilling in a copper still, with an earthenware, or pure silver, worm condenser The residues from the several distillations are collected together and redistilled to recover as much as possible of the acetic acid in them The final tarry residues are burned.

By distilling the pyroligneous acid, without neutralization, in a copper still, most of the methyl alcohol passes over before the acetic acid, by collecting the distillate until its specific gravity is about 1 000, a crude " wood spirit " is separated If the acetic acid vapors following are passed into a solution of soda, a solution of sodium acetate is obtained This is evaporated until only a fused mass of sodium acetate remains, which is heated to nearly 300° C, at which point the sodium acetate is stable but the impurities are decomposed The fused salt is dissolved in water, the solution filtered and evaporated to crystallize The process may be repeated for further purification.

If the distillation of the pyrohgneous acid be continued after the methyl alcohol has passed over, the distillate, collected between 100° and 120° C, is called " wood vinegar ", it is dilute and still retains some empyreumatic matter, but is somewhat employed technically It is generally neutralized with lime or soda to yield acetates.

Glacial acetic acid is the nearly anhydrous, 99 to 100 per cent acid, which crystallizes if cooled to 16 5° C It may be made from fused sodium acetate, by distilling with strong sulphunc acid at 120° C The residue from the still is sodium sulphate and may be used to decompose calcium acetate in solution, to prepare more sodium acetate.

Common acetic acid is found in commerce as a slightly coloured liquid of various strengths ordinary No 8 has a specific gravity about 1 040 (8° Tw) and contains approximately 30 per cent anhydrous acid, it is used in preparing acetates, in making white lead, in textile work, and in pharmacy Stronger acid, containing 50 per cent or more of anhydrous acid, is used in preparing coal-tar colours and calico printers' pastes, for prepanng organic acetates, and for solvent purposes Pure acetic acid from wood distillate may be used for vinegar, but lacks the characteristic salts and flavouring substances present in true fermentation vinegar

Acetates – Aluminum acetate in the pure state is not known, but a solution of it in acetic acid, called red liquor is largely used in dyeing and in calico printing It is made by dissolving aluminum hydroxide in acetic acid, or by decomposing lead or calcium acetates with aluminum sulphate or alum:

$$A1_2(SO_4)_3 + 3\ Pb(C_2H_3O_2)_2 = 2\ A1(C_2H_3O_2)_3 + 3\ PbSO_4$$

Calcium acetate yields the best red liquor, that made from lead acetate is not entirely free from lead, which dulls the shade of delicate colours, when made from alum, it contains sulphate of the alkali metal, and decomposes more readily than when made from aluminum sulphate Several basic aluminum acetates are made by adding sodium carbonate to the normal acetate solution These deposit alumina on the fibre very readily.

Chromium acetate finds some use as a mordant in calico printing It is usually made by dissolving chromium hydroxide in acetic acid, or by decomposing a solution of chromium sulphate or chrome alum with lead or calcium acetate The solution is violet, but becomes green if heated It may be evaporated to dryness without rendering the salt insoluble. Alkalies and alkaline carbonates yield no precipitate in the cold solution, but when heated, a precipitate of chromium hydroxide forms.

Basic acetates are prepared by adding lead or calcium acetate to basic chromium sulphate solution. Sulphate-acetates are also made and used as mordants.

Calcium acetate has been mentioned as brown or gray acetate of lime. The pure salt, occasionally used as a mordant, is made by neutralizing acetic acid with the theoretical quantity of lime. Litmus does not show the point of neutrality. The crystallized salt, $Ca(C_2H_3O_2)_2 \bullet H_2O$, is very soluble in water.

Cupric acetate, $Cu(C_2H_3O_2)_2$, H_2O, is best made by adding lead acetate to copper sulphate solution:

$$CuS0_4 + Pb(C_2H_3O_2)2 = Cu(C_2H_3O_2)_2 + PbSO_4$$

Ferrous acetate, $Fe(C_2H_3O_2)_2$, $4H_2O$, may be prepared from copperas and lead or calcium acetate, or by dissolving scrap iron in acetic acid It is quickly oxidized in the air to basic ferric acetate. " Pyrolignite of iron/" black liquor, or iron liquor, is made by dissolving scrap iron in pyrohgneous acid It is sold as a dirty olive-brown or black liquid, having a density of about 25° Tw, and consists mainly of ferrous acetate, with some ferric acetate and tarry matter It is used as a mordant in dyeing black silks and cottons, and in calico printing

Ferric acetate, $Fe(C_2H_3O_2)_3$, made by adding lead acetate to ferric sulphate, is stable m cold solution It forms basic salts when treated with caustic soda It was formerly used in black silk dyeing.

Sodium acetate, $NaC_2II_3O_2$, $3H_2O$, forms needle-like crystals which melt in their crystal water when heated, when anhydrous, it fuses without decomposition. It is chiefly used for making pure concentrated acetic acid, in making certain diazo bodies, and as a developer for the azo-dyes, m which the colour is made on the fibre

Lead acetate, $Pb(C_2H_3O_2)_2$, $3H_2O$, " sugar of lead is made by dissolving feathered lead by causing acetic acid to trickle over it in the presence of a current of air Or litharge is dissolved directly in acetic acid If wood vinegar is used, the product is " brown sugar of lead " With an excess of litharge, basic acetates are formed The normal salt is very soluble in water, and is used for making other mordants and for chrome yellows The salts are poisonous, and are affected by the carbon dioxide and hydrogen sulphide in the air.

Wood-tar varies somewhat in character with the kind of wood carbonized It is washed with hot water, or treated with milk of lime, to remove acetic acid, and then washed with very dilute sulphuric acid Excess of water is evaporated by warming in steam-jacketed vessels The tai is then distilled in iron stills, provided with stirring apparatus, the temperature being raised very slowly The distillate collected below 150° C is called " light oil/' and is chiefly used as a substitute for oil of turpentine in varnish and paints Between 150° and 250° C the " heavy oil" is collected, containing creosote, toluene, and paraffine bodies By stopping the distillation at 250° C, a thick, brownish liquid is obtained, which is used in making axle grease, shoemakers' wax, for lampblack, and for coating the interior of casks and barrels to render them impervious to liquids

The creosote oil is washed with caustic soda, and boiled in the air to oxidize various substances which it contains The alkaline solution is then acidified with sulphuric acid, to precipitate the creosote, which is treated with alkali and acid as before It is then distilled again, and the distillate, collected between 200° and 220° C, is the commercial wood-tar creosote

It has a strong, smoky odor, is a good antiseptic, and is not poisonous.

Stockholm tar and pine tar are obtained by a crude distillation of pitch-pine or other coniferous wood, in heaps, covered with turf These are of different composition from retort tar, and are mainly used for tarred ropes, with oakum for ship calking, and for preserving timber.

METALLURGY

Metallurgy is the art of extracting metals from their ores, refining them, and separating them from one another, it also includes the preparation of alloys, sometimes a careful mechanical and heat treatment of the metals is necessary in order to impart desired qualities A few metals, notably gold, platinum, silver, copper, and bismuth, are found " native," i e m the metallic state or as alloys, but generally the ores consist of oxides, sulphides, carbonates, or other salts, more or less impure, mixed with each other, with gangue rock and earthy matter.

The earliest recorded metal employed by humans appears to be gold. Small amounts of natural gold have been found in Spanish caves used during the late Paleolithic period, c. 40,000 BC. Historical developments in ferrous metallurgy can be found in a wide variety of past cultures and civilizations. This includes the ancient and medieval kingdoms and empires of the Middle East and Near East, ancient Egypt and Anatolia (Turkey), the Incas of South America, the Greeks and Romans of ancient Europe, medieval Europe, ancient and medieval China, ancient and medieval India, ancient and medieval Japan, etc. Of interest to note is that many applications, practices, and devices associated or involved in metallurgy were first established in ancient China long before Europeans mastered these crafts (such as the innovation of the blast furnace, cast iron, steel, hydraulic-powered trip hammers, etc.).

Extractive metallurgy is the practice of removing valuable metals from an ore and refining the extracted raw metals into a purer form. In order to convert a metal oxide or sulfide to a purer metal, the ore must be reduced either chemically or electrolytically.

Extractive metallurgists are interested in three primary streams: feed, concentrate (valuable metal oxide/sulfide), and tailings (waste). After mining, large pieces of the ore feed are broken through crushing and/or grinding in order to obtain particles small enough where each particle is either mostly valuable or mostly waste. Concentrating the particles of a value in a form supporting separation enables the desired metal to be removed from waste products.

Ore bodies often contain more than one valuable metal. Tailings of a previous process may be used as a feed in another process to extract a secondary product from the original ore.

Metallurgical processes form two general classes, wet and dry, the former are carried on in aqueous solution, and the latter involve changes and reactions at high temperatures

Before attempting the separation of the metal from its ore, certain preliminary operations (" ore-dressing ") are generally necessary to remove at least a part of the gangue mechanically, and to bring the valuable portion into proper condition for further treatment Sizing and concentration are accomplished by hand picking (" cob-bing") to separate large lumps, and by pulverizing and levigating the fine material The gangue, being lighter specifically than the ore, is carried off by the water The residue left from this process is called the " concentrates " If magnetic substances are present, the ore is often concentrated by allowing the pulverized material to fall between the poles of a magnet, so that the magnetic particles are deflected and separated from the non-magnetic.

In another processes, the metal is extracted by some form of leaching, generally after preliminary treatment, to get the metal into the form of a soluble salt This preliminary treatment varies much, and will be considered in connection with each special case.

In the dry processes, the ore is usually calcined or roasted and then reduced in another furnace This reduction generally consists in exposing the ore to the action of carbon and carbon monoxide at high temperatures, or sulphur may be relied upon to take oxygen from the oxides in the prepared ore The gangue substances are rendered fluid by adding fluxes which fuse to liquid slag with them. Commonly the flux is silica (SiO_2) where acid conditions are desired; for basic, lime is used.

The crude metal obtained by any reduction process generally requires "refining," to remove the impurities left in it The methods for this refining will be considered in connection with the individual metals.

ROASTING

Roasting consists in producing chemical changes at a temperature \\hich, though quite high, is not sufficient to cause fusion No metallurgical operation is more important, for the results of various processes depend on converting the ore into a suitable chemical condition by roasting For example, in the chlorination of gold ores, sulphur and arsenic must first be completely removed, or, to reduce zinc from sulphide, the latter must first be converted into oxide.

An oxidizing roast serves to remove sulphur, arsenic, etc, by conversion to SO_2, $As_2()3$, etc, which are carried away by the draft.

$$PbS + 3O = PbO + SO_2$$

A sulphatizmg roast converts sulphides to sulphates, usually with the object of leaving some metal in a soluble condition This is accomplished by keeping the temperature somewhat lower, and the depth of the ore bed greater, than for an oxidizing roast, and reducing the draft These conditions favour the maximum production from the sulphur in the ore, and prevent it being carried into the stack before it can combine with the metallic oxides.

When it is necessary to remove all of the sulphur, arsenic, etc, from an ore, any sulphates, arsenates, etc, that have formed are reduced by stirring in fine coal with the ore and excluding the air as much as possible. The resulting sulphides are then given a further oxidizing roast When the sulphur, arsenic, etc, are practically all removed, the ore is said to be dead roasted, or sweet roasted.

A chloridizing roast converts metals into chlorides by means of the interaction of atmospheric oxygen and common salt with the sulphides, arsenides, etc. In the most important case of chloridizing, that of silver, the sulphide is first oxidized to sulphate, and then the following reaction takes place.

$$Ag_2SO_4 + 2\,NaCl \rightarrow 2\,AgCl + Na_2SO_4$$

Roasting of fine ores is more common than of lump Fines are heated in beds only a few inches deep, so that they do not pack and prevent their proper exposure to the air. They are turned over and over to expose new surfaces Coarse ores worked in this way can receive only a superficial roast in any reasonable time, and are therefore roasted in large heaps 5 or 6 feet deep.

PRODUCTION ENGINEERING OF METALS

Metals are shaped by processes such as casting, forging, Flow Forming, Rolling (metalworking), extrusion, sintering, metalworking, machining and fabrication. With casting, molten metal is poured into a shaped mould. With forging, a red-hot billet is hammered into shape. With rolling, a billet is passed through successively narrower rollers to create a sheet. With extrusion, a hot and malleable metal is forced under pressure through a die, which shapes it before it cools. With sintering, a powdered metal is compressed into a die at high temperature. With machining, lathes, milling machines, and drills are used to cut the cold metal to shape. With fabrication, sheets of metal are cut with guillotines or gas cutters and bent into shape.

"Cold working" processes, where the product's shape is altered by rolling, fabrication or other processes while the product is cold, can increase the strength of the product by a process called work hardening. Work hardening creates microscopic defects in the metal, which resist further changes of shape. Various forms of casting exist in industry and academia. These include sand casting, investment casting (also called the "lost wax process"), die casting and continuous casting.

Welding is a technique for joining certain ferrous metals and certain aluminium alloys. The metals in the weld and on both sides of the join are generally similar alloys. Brazing is a technique for joining ferrous or non-ferrous metals with a copper-based (generally brass or bronze) filler.

Metals can be heat-treated by annealing, quenching, tempering and case hardening to alter properties of toughness, hardness or resistance to corrosion. Annealing softens the metal and makes a shaped product tougher by reducing the effects of work hardening. Quenching and case hardening are used to make a shaped product harder. Quenching by itself makes the metal very hard and very brittle. Tempering after quenching is used to reduce the brittleness and improve overall properties.

Electroplating is the main surface-treatment technique. It involves bonding a thin layer of another metal such as gold, silver, chromium or zinc to the surface of the product. It is used to reduce corrosion as well as to improve the product's aesthetic appearance.

Metallurgists study the microscopic and macroscopic mechanisms that cause a metal or alloy to behave in the way that it does, i.e. the changes that occur on the atomic level that affect the metal's (or alloy's) macroscopic properties. Examples of tools used for microscopic examination of metals are optical and electron microscopes and mass spectrometres.

Metallurgists study crystallography, the effects of temperature and heat treatment on the component phases of alloys, such as the eutectic and the properties of those alloy phases. The macroscopic properties of metals are tested using machines and devices that measure tensile strength, compressive strength and hardness.

